Collins

KS3
Maths
Higher Level

Revision Guide

Samya Abdullah, Rebecca Evans,
Trevor Senior and Gillian Spragg

About this Revision & Practice book

When it comes to getting the best results, practice really does make perfect!

Experts have proved that repeatedly testing yourself on a topic is far more effective than re-reading information over and over again. And, to be as effective as possible, you should space out the practice test sessions over time.

This Complete Revision & Practice book is specially designed to support this approach to revision and includes seven different opportunities to test yourself on each topic, spaced out over time.

This book is suitable for students working at or above the expected level in Key Stage 3 Maths.

Symbols are used to highlight questions that test key skills:

(MR) Mathematical Reasoning   (PS) Problem Solving

(FS) Financial Skills

Try to answer as many questions as possible without using a calculator.

Questions where calculators **must not** be used are marked with this symbol:

Revise

These pages provide a recap of everything you need to know for each topic. All key words are defined in the glossary.

You should read through all the information before taking the Quick Test at the end. This will test whether you can recall the key facts.

> **Quick Test**
>
> 1. Simplify $4x + 7y + 3x - 2y + 6$
> 2. Simplify $c \times c \times d \times d$
> 3. Find the value of $4x + 2y$ when $x = 2$ and $y = 3$

Practise

These topic-based questions appear shortly after the revision pages for each topic and will test whether you have understood the topic. There are two levels of demand for each topic.

Review

These topic-based questions appear later in the book, allowing you to revisit the topic and test how well you have remembered the information. There are two levels of demand for each topic.

Mix it Up

These pages feature a mix of questions for all the different topics, just like you would get in a test. They will make sure you can recall the relevant information to answer a question without being told which topic it relates to.

Test Yourself on the Go

Visit our website at **collins.co.uk/collinsks3revision** and print off a set of flashcards. These pocket-sized cards feature questions and answers so that you can test yourself on all the key facts anytime and anywhere. You will also find lots more information about the advantages of spaced practice and how to plan for it.

Workbook

This section features even more topic-based questions (again with two levels of demand) and mixed test-style questions, providing two further practice opportunities for each topic to guarantee the best results.

ebook

To access the ebook, visit

collins.co.uk/ebooks

and follow the step-by-step instructions.

QR Codes

Found throughout the book, the QR codes can be scanned on your smartphone for extra practice and explanations.

A QR code in the Revise section links to a Quick Recall Quiz on that topic. A QR code in the Workbook section links to a video working through the solution to one of the questions on that topic.

Contents

N Number   **A** Algebra   **G** Geometry and Measures

S Statistics   **P** Probability   **R** Ratio, Proportion and Rates of change

Contents   3

Contents

N Number **A** Algebra **G** Geometry and Measures
S Statistics **P** Probability **R** Ratio, Proportion and Rates of change

Contents

Review Questions

Key Stage 2: Key Concepts

1 Which of these two numbers is closer to 2000?

1996 or 2007

Explain how you know. [2]

2 Calculate 476 − 231  [2]

3 Copy and complete the table below by rounding each number to the nearest 1000

	To the nearest 1000
4587	
45 698	
457 658	
45 669	

[2]

4 Write these values in order, starting with the smallest.

0.56   55% $\frac{27}{50}$ 0.6   0.63 [3]

5 Ahmed is twice as old as Rebecca.

Rebecca is three years younger than John.

John is 25 years old.

How old is Ahmed? [2]

6 Calculate 467 × 34 [2]

7 Calculate 156 ÷ 3 [2]

8 On the scale below, draw arrows to show 1.6 and 3.8

[2]

Total Marks _____ / 17

1 A bottle holds 1 litre of fizzy drink. Mariam pours four glasses for her friends.

Each glass contains 200 ml.

How much fizzy drink is left in the bottle? [3]

2 Below are five digit cards.

| 7 | 5 | 1 | 6 | 3 |

Choose two cards to make the following two-digit numbers.

a) A square number [1]

b) A prime number [1]

c) A multiple of 6 [1]

d) A factor of 60 [1]

3 An equilateral triangle has a perimeter of 27 cm.

What is the length of one of its sides? [2]

4 Two-thirds of a number is 22

What is the number? [2]

5 Here is an isosceles triangle drawn inside a rectangle.

Find the value of the angle x. [3]

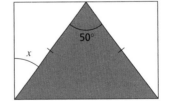

6 S and T are two whole numbers.

$S + T = 500$

S is 100 greater than T.

Find the value of S and T. [2]

Total Marks .... / 16

Number 1

You must be able to:

- Carry out calculations with negative numbers
- Use the symbols $=, \neq, <, >, \leq, \geq$
- Multiply and divide integers
- Carry out operations following BIDMAS.

Comparing and Working with Numbers

- An **integer** is a whole number.
- You can use a number line to add and subtract numbers.

$$2 - 6 = -4$$

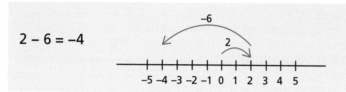

> **Key Point**
>
> There is an infinite number of positive and negative numbers.

- **Place value** can be used to compare the size of large numbers.

Which is greater, 3408 or 3540?

Number	Thousands	Hundreds	Tens	Units
3408	3	4	0	8
3540	3	5	4	0

Both numbers have 3 thousands, but 3540 has 5 hundreds and 3408 only has 4 hundreds. Therefore 3540 is greater than 3408.

> **Key Point**
>
> Always compare digits from left to right.

- **Symbols** are used to state the relationship between two numbers.

Symbol	Meaning	Symbol	Meaning
$=$	Equal to	$\neq$	Not equal to
$<$	Less than	$\leq$	Less than or equal to
$>$	Greater than	$\geq$	Greater than or equal to

3540 is greater than 3408 can be written as $3540 > 3408$
−10 is less than −5 can be written as $-10 < -5$
$-6 < -2$ $3 > -1$ $2 + 3 \neq 23$

- The table below shows the rules to use when carrying out calculations with negative numbers.
- When adding and subtracting negative numbers, the rules only apply if the signs are next to each other.

	+	−
+	+	−
−	−	+

$-5 \times -3 = +15$
$-20 \div 4 = -5$
$-6 - (+5) = -6 - 5 = -11$
$-6 + 9 = +3$

Multiplication and Division

- To multiply large numbers, you can use the **grid** (or **box**) method or a **column** method.

Calculate 354×273

Grid or Box method

$354 = 300 + 50 + 4$
$273 = 200 + 70 + 3$

×	300	50	4	
200	60 000	10 000	800	70 800
70	21 000	3500	280	24 780
3	900	150	12	1 062
				96 642

Adding the numbers in the grid gives $354 \times 273 = 96\,642$

Column method

```
        3   5   4
    ×   2   7   3
    1   0¹  6¹  2     (3 × 354)
  2 4³  7²  8   0     (70 × 354)
  7 0   8   0   0     (200 × 354)
  9 6¹  6¹  4   2
```

So, $354 \times 273 = 96\,642$

- Division can also be broken down into steps.

Calculate $762 \div 3$

Short division

$3\overline{)7\,6\,2}$	$3\overline{)7\,{}^16\,2}$ $\,{}^2$	$3\overline{)7\,{}^16\,{}^12}$ $\,{}^{2\,5}$	$3\overline{)7\,{}^16\,{}^12}$ $\,{}^{2\,5\,4}$
Set up the division.	$7 \div 3 = 2$ remainder 1	$16 \div 3 = 5$ remainder 1	$12 \div 3 = 4$ remainder 0

So, $762 \div 3 = 254$

Long division

```
      2 5 4
  3 ) 7 6 2
      6
      1 6
      1 5
        1 2
        1 2
          0
```

Key Point

Division and multiplication should be carried out in the order they appear in the calculation from the left. So carry out the multiplication first if it appears before the division.

Addition and subtraction should be carried out in the order they appear in the calculation from the left. So carry out the subtraction first if it appears before the addition.

BIDMAS

- **BIDMAS** gives the order in which operations should be carried out:

  Brackets
  Indices (powers)
  Division *and*
  Multiplication
  Addition *and*
  Subtraction

$4 \times 5 + 6^2 \div 4$
$= 4 \times 5 + 36 \div 4$
$= 20 + 9$
$= 29$

$4 \times 5 + 6^2 \div 4 + 0$
$= 4 \times 5 + 36 \div 4 + 0$
$= 20 + 9 + 0$
$= 29$

Adding 0 does not change the answer. 0 is called an **additive identity** as it leaves the answer unchanged.

$(3 + 4) \times (9 - 1) = 7 \times 8 = 56$
$(3 + 4) \times (9 - 1) \times 1 = 7 \times 8 \times 1 = 56$

Multiplying by 1 does not change the answer. 1 is called a **multiplicative identity** as it leaves the answer unchanged.

Key Words

integer
place value
BIDMAS
additive identity
multiplicative identity

Quick Test

1. Work out -5×-7
2. Work out 435×521
3. Work out $652 \div 4$
4. Work out $4 \times 3^2 + 7 \times 4$

Number 2

You must be able to:

- Understand square numbers, square roots, cube numbers and cube roots
- Write a number as a product of prime factors
- Find the lowest common multiple and highest common factor.

Squares, Square Roots, Cubes and Cube Roots

- **Square numbers** are calculated by multiplying a number by itself.

  $$5^2 = 5 \times 5 = 25$$

- A **square root** ($\sqrt{}$) is the inverse or opposite of a square.

  $\sqrt{36} = 6$ (because $6 \times 6 = 36$) $\sqrt{25} = 5$ (because $5 \times 5 = 25$)

  $\sqrt{30}$ is between 5 and 6 $\sqrt{30} = 5.47...$

- **Cube numbers** are calculated by multiplying a number by itself and by itself again.

  $$4^3 = 4 \times 4 \times 4 = 64 \qquad 4^2 + 3^3 = 16 + 27 = 43$$

- A **cube root** ($\sqrt[3]{}$) is the inverse or opposite of a cube.

  $$\sqrt[3]{8} = 2 \text{ because } 2 \times 2 \times 2 = 8$$

- There are other powers as well as squares and cubes.

  $$5^4 = 5 \times 5 \times 5 \times 5 = 625 \qquad 5^5 = 5 \times 5 \times 5 \times 5 \times 5 = 3125$$

> **Key Point**
>
> The first ten square numbers are:
>
> 1, 4, 9, 16, 25, 36, 49, 64, 81, 100

> **Key Point**
>
> The first five cube numbers are:
>
> 1, 8, 27, 64, 125

Reciprocals

- A **reciprocal** is the inverse of any number except zero, e.g. the reciprocal of 2 is $\frac{1}{2}$

The reciprocal of $\frac{1}{2}$ is $1 \div \frac{1}{2} = 2$

Prime Factors

- A **factor** of an integer is an integer that divides exactly into it without leaving a remainder. Factors are integers you can multiply together to make another number.

  The factors of 12 are 1, 2, 3, 4, 6 and 12 because $1 \times 12 = 12$, $2 \times 6 = 12$ and $3 \times 4 = 12$

- Every integer greater than 1 is either prime or can be written as the **product** of a unique list of prime numbers.
- A **prime** number has exactly two factors, itself and 1

This is called the unique factorisation property.

> **Key Point**
>
> Product means multiply.

- Breaking a number down into a product of prime factors is called **prime factor decomposition**.

45 can be expressed as a product of prime factors. This can be done using a **prime factor tree**:

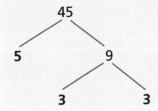

$45 = 5 \times 9$
and
$9 = 3 \times 3$
So $45 = 3 \times 3 \times 5$

LCM and HCF

- The **lowest common multiple (LCM)** is the lowest multiple two or more numbers have in common.
- The **highest common factor (HCF)** is the highest factor two or more numbers have in common.

Find the lowest common multiple and highest common factor of 12 and 42

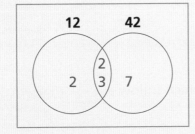

The LCM is the product of all the numbers in both circles.

LCM $= 2 \times 2 \times 3 \times 7 = 84$

The HCF is the product of the numbers in the overlap.

HCF $= 2 \times 3 = 6$

Write both numbers as a product of prime factors.
$12 = 2 \times 2 \times 3$ $42 = 2 \times 3 \times 7$
Complete a Venn diagram.

Common factors are placed in the overlap.

- LCM and HCF are used to solve many everyday problems.

Tom swims in competitions. He visits his doctor every 12 days. He visits his nutritionist every 15 days. He saw both of them on 1st October. On what date will he next see both of them on the same day?

Find the LCM of 12 and 15: this is 60

60 days after 1st October is 30th November.

Quick Test

1. Write down the values of: **a)** 7^2 **b)** 4^3
2. Write down the values of: **a)** $\sqrt{49}$ **b)** $\sqrt[3]{27}$
3. Write 40 as a product of prime factors.
4. Find the lowest common multiple of 14 and 36
5. Find the highest common factor of 24 and 32

Sequences 1

You must be able to:

- Use a function machine to generate terms of a sequence
- Recognise arithmetic and geometric sequences
- Generate sequences from a term-to-term rule.

Function Machines

- A **function machine** (number machine) takes an input, applies one or more operations and produces an output.
- Function machines can be shown in different ways:

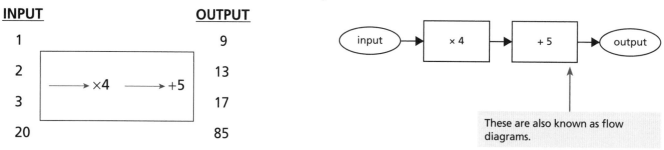

These are also known as flow diagrams.

Sequences

- A **sequence** is a set of shapes, numbers or letters which follow a pattern or rule.

The outputs from a function machine can form a sequence.

- Each part of a sequence is called a **term**.
- Any sequence that can carry on forever is called **infinite**.

In the sequence below, the next pattern is formed by adding extra tiles around the previous pattern.

| Basic design | Layer 1
6 new tiles | Layer 2
10 new tiles | Layer 3
14 new tiles |

The number of new tiles needed increases by 4

This pattern can be used to predict how many tiles will be needed to make larger designs.

- An **arithmetic sequence** is a set of numbers with a **common difference** between consecutive terms.

> 4, 7, 10, 13, 16,… is an arithmetic sequence with a common difference of 3
>
> 12, 8, 4, 0, −4,… is also an arithmetic sequence. It has a common difference of −4

- In a **geometric sequence**, each term is multiplied by the same value to find the next term.

Key Point

Arithmetic sequences are used to solve many real-life problems.

2, 4, 8, 16, 32, 64,... is a geometric sequence where each term is multiplied by 2 to find the next term.

12, 6, 3, 1.5,... is also a geometric sequence where the previous term is multiplied by $\frac{1}{2}$ to find the next term.

- Many other number sequences follow patterns:
  - 1, 4, 9, 16, 25, 36,... are square numbers.
  - 1, 8, 27, 64, 125, 216,... are cube numbers.
  - 1, 3, 6, 10, 15, 21,... are known as the **triangular numbers**.
  - 1, 1, 2, 3, 5, 8, 13,... is known as the **Fibonacci sequence**.

- When calculating terms of a geometric sequence, it is quicker to use powers.

  A culture of bacteria doubles every 4 hours. There are 1000 bacteria at the beginning. How many will there be in 12 hours?

  $12 \div 4 = 3$, so it will double three times.

  $1000 \times 2^3 = 8000$ bacteria

> **Key Point**
>
> There are other types of sequences, for example exponentials and reciprocals.

Each number is the sum of the previous two numbers.

Remember, $x^4 = x \times x \times x \times x$
$x^5 = x \times x \times x \times x \times x$

$2^3 = 2 \times 2 \times 2$

Finding Missing Terms

- The **term-to-term rule** links each term in the sequence to the previous term.

  $5, 6\frac{1}{3}, 7\frac{2}{3}, 9, 10\frac{1}{3},...$

  In this set of numbers the next term is found by adding $1\frac{1}{3}$ to the previous term. Therefore the term-to-term rule is $+1\frac{1}{3}$

  This rule can be used to find the next numbers in the sequence:

  $10\frac{1}{3} + 1\frac{1}{3} = 11\frac{2}{3}$ $11\frac{2}{3} + 1\frac{1}{3} = 13$ $13 + 1\frac{1}{3} = 14\frac{1}{3}$

  Therefore the next three terms in the sequence are

  $11\frac{2}{3}, 13, 14\frac{1}{3}$

- The term-to-term rule can also be used to find missing terms.

  13, 8, 3, _____, –7,...

  The term-to-term rule is –5 and so the missing term is –2

> **Key Words**
>
> function machine
> sequence
> term
> infinite
> arithmetic sequence
> common difference
> geometric sequence
> triangular numbers
> term-to-term

> **Quick Test**
>
> 1. Write down the term-to-term rule for this sequence:
>   24, 12, 6, 3, 1.5,...
> 2. Which of these sequences is arithmetic?
>   4, 6, 9, 13, 18,...   5, 9, 13, 17, 21,...
> 3. Find the missing term in the following sequence of numbers.
>   15, 9, 3, _____, –9, –15,...
> 4. Which of these sequences is geometric?
>   4, 8, 16, 32, 64,...   16, 14, 12, 10, 8,...

Sequences 2

You must be able to:

- Generate terms of a sequence from a position-to-term rule
- Find the nth term of an arithmetic sequence
- Recognise quadratic sequences.

The nth Term

- The **nth term** is also called the **position-to-term rule**.
- It is an algebraic expression that represents the operations carried out by a function machine.

INPUT		OUTPUT
1		9
2	$\longrightarrow \times 4 \longrightarrow + 5$	13
3		17
n		$4n + 5$

- The **nth term** can be used to generate terms of a sequence.

The nth term of a sequence is given by $3n + 5$

To find the first term, you **substitute** $n = 1$

$3 \times 1 + 5 = 8$ 8 is the first term in the sequence.

To find other terms, you can substitute different values of n.

When $n = 2$	When $n = 3$	When $n = 4$
$3 \times 2 + 5 = 11$	$3 \times 3 + 5 = 14$	$3 \times 4 + 5 = 17$
Second term = 11	Third term = 14	Fourth term = 17

The nth term $3n + 5$ produces the sequence of numbers:

8, 11, 14, 17, 20,...

The rule can be used to find any term in the sequence.
For example, to find the 50th term in the sequence
substitute $n = 50$

$3 \times 50 + 5 = 155$

> **Key Point**
>
> For the first term in the sequence, n always equals 1

The nth term of a sequence is given by $13 - 4n$

Work out the first four terms of the sequence.

Substituting $n = 1, 2, 3, 4$ gives 9, 5, 1, −3

In this sequence the terms are decreasing.

Finding the nth Term

- To find the nth term, look for a pattern in the sequence of numbers.

The first five terms of a sequence are 7, 11, 15, 19, 23

The term-to-term rule is +4 so the nth term starts with $4n$.

The difference between $4n$ and the output in each case is 3, so the final rule is $4n + 3$

Input	× 4	Output
1	4	7
2	8	11
3	12	15
4	16	19
5	20	23
n	$4n$	$4n + 3$

Quadratic Sequences

- **Quadratic** sequences are based on square numbers.

The first five terms of the sequence $2n^2 + 1$ are:
3, 9, 19, 33, 51,...

The nth term of a sequence is given by $\dfrac{n(n+1)}{2}$

Work out the first four terms of the sequence.

Substituting $n = 1, 2, 3, 4$ gives 1, 3, 6, 10

Key Point

Use BIDMAS when calculating terms in a sequence.

This is the sequence of triangular numbers.

- Triangular numbers are produced from a quadratic sequence.
- There are many other sequences, for example the first five terms for n^3 are 1, 8, 27, 64, 125

Quick Test

1. Write down the first five terms in the sequence $5n + 3$
2. Write down the first five terms in the sequence $5n^2 - 1$
3. a) Find the nth term for the following sequence of numbers:
   20, 16, 12, 8, 4,...
   b) Find the 50th term in this sequence.
4. What is the nth term also known as?

Key Words

nth term
position-to-term
substitute
quadratic

Practice Questions

Number

(MR) 1 Jessa and Holly have been given the following question:

What is the value of $3 + 5 \times 4 + 7$?

Jessa thinks the answer is 30 and Holly thinks the answer is 39. Who is right?
Explain your answer. [2]

(FS) 2 A netball club is planning a trip. The club has 354 members and the cost of the trip is £12 per member.

    a) Work out the total cost of the trip. [3]

    They need coaches for the trip and each coach seats 52 people.

    b) How many coaches do they need to book? [3]

    c) How many spare seats will there be? [2]

(FS) 3 Alicia wants to buy 15 hotdogs.

Dave's dogs	**Harry's hotdogs**
3 hotdogs £4.00	5 hotdogs £6.00

Is Dave's dogs or Harry's hotdogs cheaper? Show your working. [3]

4 Fill in the missing number.

$$2 - \boxed{} = 5$$

[1]

Total Marks .... / 14

---

1 120 can be written in the form $2^p \times q \times r$ where p, q and r are prime numbers.

Find the value of each of p, q and r. [3]

2 x and y are two different prime numbers.

Find the highest common factor of the two expressions x^3y and xy^3 [1]

Total Marks .... / 4

Sequences

1. a) Find the nth term of this arithmetic sequence: 4, 7, 10, 13, 16,… [3]

   b) Find the 60th term in the sequence. [1]

(MR) 2. Match the cards on the left with the correct card on the right.

5, 9, 13, 17, 21,…	Neither
2, 8, 18, 32, 50,…	Quadratic
8, 17, 32, 53, 80,…	Arithmetic

   [2]

(MR) 3. a) Explain why $\sqrt{79}$ must be between 8 and 9 [2]

   b) Use your calculator to find the value of $\sqrt{79}$ to 2 decimal places. [1]

Total Marks .... / 9

1. Match the sequence of numbers with the correct nth term.

2, 7, 12, 17, 22,…	$5 - n$
3, 9, 27, 81,…	$5n^2 + 1$
6, 21, 46, 81, 126,…	$5n - 3$
4, 3, 2, 1, 0,…	3^n

   [2]

(PS) 2. The half-life of a radioactive material is the time taken for the level of radioactivity to decrease to half of its initial level.

A radioactive material is found which has a half-life of 1 day. The initial level in the sample taken was 800 units.

Find the amount of radioactive material left in the sample at the end of the fifth day. [2]

Total Marks .... / 4

Perimeter and Area 1

You must be able to:

- Find the perimeter and area of a square
- Find the perimeter and area of a rectangle
- Find the area of a triangle
- Find the area and perimeter of compound shapes.

Perimeter and Area of Squares and Rectangles

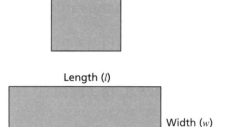

Length (*l*)

Length (*l*)

Width (*w*)

- The **perimeter** is the distance around the outside of a 2D shape.
- The formula for the perimeter of a square is $4 \times$ length or $P = 4l$
- The formula for the perimeter of a rectangle is:
  perimeter = 2(length + width) or $P = 2(l + w)$
  also perimeter = 2(length) + 2(width) or $P = 2l + 2w$
- The formula for the **area** of a square or rectangle is:
  area = length $\times$ width or $A = l \times w$

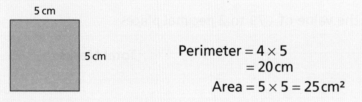

Find the perimeter and area of this square.

5 cm

5 cm

Perimeter $= 4 \times 5$
$= 20$ cm
Area $= 5 \times 5 = 25$ cm²

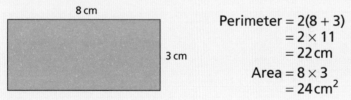

Find the perimeter and area of this rectangle.

8 cm

3 cm

Perimeter $= 2(8 + 3)$
$= 2 \times 11$
$= 22$ cm
Area $= 8 \times 3$
$= 24$ cm²

Area of a Triangle

- The formula for the area of a triangle is:
  area $= \frac{1}{2}$(base $\times$ **perpendicular** height)

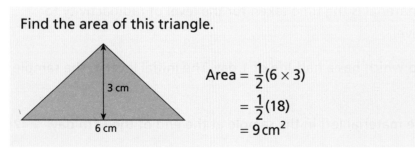

Find the area of this triangle.

3 cm

6 cm

Area $= \frac{1}{2}(6 \times 3)$
$= \frac{1}{2}(18)$
$= 9$ cm²

- The **altitude of a triangle** is a perpendicular line drawn from a vertex to the opposite side.

> **Key Point**
>
> When finding the area of a triangle, always use the perpendicular height.

Area and Perimeter of Composite Shapes

- A **composite** shape is made up from other, simpler shapes.
- To find the area of a composite shape, divide it into basic shapes.

This shape can be broken up into three rectangles.

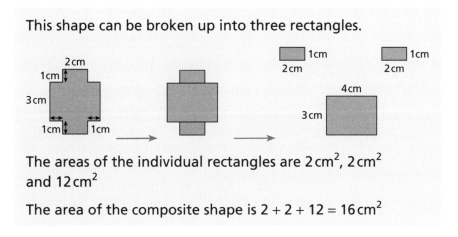
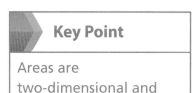

The areas of the individual rectangles are $2\,cm^2$, $2\,cm^2$ and $12\,cm^2$

The area of the composite shape is $2 + 2 + 12 = 16\,cm^2$

- To find the perimeter, start at one corner of the shape and travel around the outside, adding the lengths.

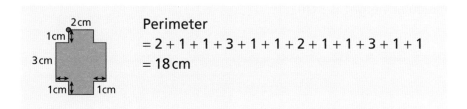

Perimeter
$= 2 + 1 + 1 + 3 + 1 + 1 + 2 + 1 + 1 + 3 + 1 + 1$
$= 18\,cm$

- Shapes that have straight edges and right angles are called **rectilinear** shapes.

Quick Test

1. Find the perimeter of a rectangle with width 5 cm and length 7 cm.
2. Find the area of a rectangle with width 9 cm and length 3 cm. Give appropriate units in your answer.
3. Find the area of a triangle with base 4 cm and perpendicular height 3 cm.
4. Find the perimeter and area of this shape.

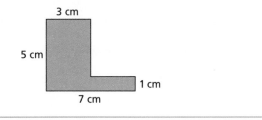

Perimeter and Area 2

You must be able to:

- Find the area of a parallelogram
- Find the area of a trapezium
- Find the circumference and area of a circle.

Area of a Parallelogram

- A **parallelogram** has two pairs of **parallel** sides.
- The formula for the area of a parallelogram is:

area = base × perpendicular height

Find the area of this parallelogram.

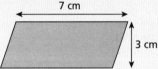

7 cm

3 cm

The base is 7 cm and the perpendicular height is 3 cm.

Area = 7 × 3

= 21 cm^2

Key Point

Parallel lines travel in the same direction. They stay the same distance apart and never meet.

Area of a Trapezium

- A **trapezium** has one pair of parallel sides.
- The formula for the area of a trapezium is:

$A = \frac{1}{2}(a + b)h$

- The sides labelled **a** and **b** are the parallel sides and **h** is the perpendicular height.
- Perpendicular means at right angles.

Find the area of this trapezium.

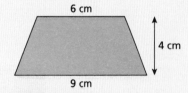

6 cm

4 cm

9 cm

Area = $\frac{1}{2}$(6 + 9) × 4

= 30 cm^2

Circumference and Area of a Circle

- The **circumference** of a **circle** is the distance around the outside.
- The **radius** of a circle is the distance from the centre to the circumference. The **diameter** of a circle is twice the radius.
- The formula for the circumference of a circle is: $C = 2\pi r$ or $C = \pi d$
- The formula for the area of a circle is: $A = \pi r^2$

Find the circumference and area of this circle. Give your answers to 1 decimal place.

7 cm

The circumference:
$$C = 2 \times \pi \times 7$$
$$= 14 \times \pi$$
$$= 44.0 \, \text{cm}$$

The area:
$$A = \pi \times 7^2$$
$$= \pi \times 49$$
$$= 153.9 \, \text{cm}^2$$

- Circles can be split into **sectors**. A sector is a region bounded by two radii and an **arc** (a curved line that is part of the circumference).
- Arc lengths and areas of sectors can be calculated using fractions of $360°$

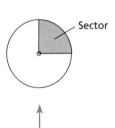
Sector

Find the perimeter and area of the sector of a circle. Give your answers to 1 decimal place.

$60°$ is one-sixth of $360°$

60°
4 cm

Arc length $= \dfrac{1}{6} \times 2 \times \pi \times 4 = 4.2 \, \text{cm}$

Perimeter $= 4.2 + 4 + 4 = 12.2 \, \text{cm}$

Area $= \dfrac{1}{6} \times \pi \times 4^2 = 8.4 \, \text{cm}^2$

A sector with a 90° angle at the centre would have an area of $\frac{1}{4}$ of the whole circle.

Find the shaded region of a circle with radius 5 cm when the angle at the centre is $30°$

5 cm
30°

The area of the whole circle is $\pi \times 5^2 = 25\pi$

The shaded sector is $\dfrac{30}{360}$ th of the circle $= \dfrac{1}{12}$ th

Area of sector $= \dfrac{25\pi}{12} = 6.5 \, \text{cm}^2$ (1 d.p.)

Quick Test

1. Find the area of the parallelogram.

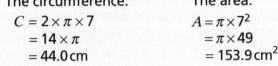
8 cm
2 cm

2. Find the area of the trapezium.

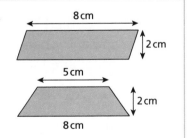
5 cm
2 cm
8 cm

3. Find the circumference and area of a circle with diameter 6 cm.
4. Find the area of a sector with radius 4 cm and angle $40°$ at the centre.

Statistics and Data 1

You must be able to:

- Find the mean, median, mode and range for a set of data
- Choose which average is the most appropriate to use in different situations
- Use a tally chart to collect data
- Construct bar charts and vertical line graphs.

Mean, Median, Mode and Range

- The **mean** is the **sum** of all the values divided by the number of values.
- The **median** is the middle value when the data is in order.
- The **mode** is the most common value.
- The mean, median and mode are all averages and are called **measures of central tendency**.
- The **range** is the **difference** between the biggest and the smallest value.
- The range is a measure of spread or a **measure of dispersion**.

> **Key Point**
>
> Data can have more than one mode. Bi-modal means the data set has two modes.

Find the mean, median, mode and range for this data:

$$5, 9, 7, 9, 2, 7, 3, 9, 4, 2, 6, 6, 4, 5$$

$$\text{The mean} = \frac{5+9+7+9+2+7+3+9+4+2+6+6+4+5}{14}$$

$$= 5.6 \text{ (to 1 d.p.)}$$

The median = 2, 2, 3, 4, 4, 5, ⑤⑥ 6, 7, 7, 9, 9, 9

The median is the midpoint of 5 and 6 so 5.5

The mode is 9 as this number is seen most often.

The range = 9 − 2 = 7

← Put the data in order, smallest to biggest.

Choosing an Appropriate Average

- Use the **mode** when you are interested in the most common answer, for example if you were a shoe manufacturer deciding how many of each size to make.
- Use the **mean** when your data does not contain **outliers**. A company which wanted to find average sales across a year would want to use all values.
- Use the **median** when your data does contain outliers, for example finding the average salary for a company when the manager earns many times more than the other employees.

> **Key Point**
>
> An outlier is a value that is much higher or lower than the others.

Constructing a Tally Chart

- A **tally chart** is a quick way of recording data.
- Your data is placed into groups, which makes it easier to analyse.
- A tally chart can be used to make a frequency table by adding an extra column to record the total in each group.

> **Key Point**
>
> The frequency is the total for the group.

You are collecting and recording data about people's favourite flavour crisps. You ask 50 people and fill in the tally chart with their responses.

Flavour	Tally	Frequency
Plain	ⵕ ⵕ II	12
Salt and vinegar	ⵕ IIII	9
Cheese and onion	ⵕ ⵕ ⵕ I	16
Prawn cocktail	ⵕ II	7
Other	ⵕ I	6
Total		50

From the table, you can see that cheese and onion is the mode.

Bar Charts and Vertical Line Graphs

* **Bar charts** and **vertical line graphs** can be used to compare frequencies.

Draw a bar chart and a vertical line graph to represent the data above.

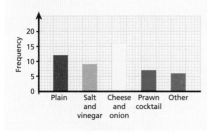 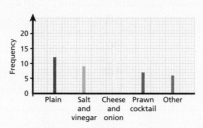

Quick Test

1. Emma surveyed her class to find out their favourite colour. She constructed this tally chart.
   a) Complete the frequency column.
   b) How many students are there in Emma's class?

Colour	Tally	Frequency
Red	ⵕ IIII	
Blue	ⵕ II	
Green	ⵕ III	
Yellow	ⵕ I	
Other	ⵕ II	

2. Look at this set of data:
   3, 7, 4, 6, 3, 5, 9, 40, 6
   a) Write down the mode.
   b) Calculate the mean, median and range.
   c) Would you choose the mean or the median to represent this data? Explain your answer.

Key Words

mean
sum
median
mode
range
difference
outlier
tally chart
bar chart
vertical line graph

Statistics and Data 2

You must be able to:

* Group data and construct grouped frequency tables
* Construct and interpret a two-way table.

Grouping Data

* When you have a large amount of data, it is sometimes appropriate to place it into groups.
* A group is also called a **class interval**.
* The disadvantage of using **grouped data** is that the original **raw data** is lost.

> | Key Point |
> |---|
> | Calculations based on grouped data will be estimates. |

The data below represents the number of people who visited the library each day over a 60-day period.

Number of people	Frequency
0–10	10
11–20	30
21–30	14
31–40	6

On 30 out of the 60 days, the library had between 11 and 20 (inclusive) visitors.

* This information can be shown using a bar chart (see right).
* To estimate the mean of grouped data, the midpoint of each class is used.

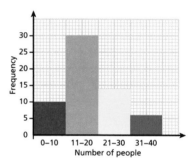

Calculate an estimate of the mean for the data above.

> | Key Point |
> |---|
> | Data such as the number of people is discrete as it can only take particular values. Data such as height and weight is continuous as it can take any value on a particular scale. |

Number of people	Midpoint (x)	Frequency (f)	fx
0–10	5	10	50
11–20	15.5	30	465
21–30	25.5	14	357
31–40	35.5	6	213
Total		**60**	**1085**

Estimate of mean is

$$\frac{50 + 465 + 357 + 213}{60} = \frac{1085}{60}$$
$$= 18.1 \text{ (to 1 d.p.)}$$

Two-Way Tables

- A two-way table shows information that relates to two different categories.
- Two-way tables can be constructed from information collected in a survey.

Daniel surveyed his class to find out if they owned any pets. In his class there are 16 boys and 18 girls. 10 of the boys owned a pet and 15 of the girls owned a pet.

	Pets	No Pets	Total
Boys	10		16
Girls	15		18
Total			

The information given is filled into the table and then the missing information can be worked out.

If there are 16 boys in Daniel's class and 10 of them have pets, then to work out how many boys do not have pets we calculate $16 - 10 = 6$

When all the missing information is entered you can **interpret** the data.

	Pets	No Pets	Total
Boys	10	6	16
Girls	15	3	18
Total	25	9	34

You can see that there are 34 students in Daniel's class and 25 of them owned a pet. You can also see that more girls owned pets than boys, and that more students owned pets than did not own pets.

Quick Test

1. 25 women and 30 men were asked if they preferred football or rugby. 16 of the women said they prefer football and 10 of the men said they prefer rugby.
   a) Construct a two-way table to represent this information.
   b) How many in total said they prefer football?
   c) How many women preferred rugby?
   d) How many people took part in the survey?

> **Key Words**
>
> class interval
> grouped data
> raw data
> interpret

Review Questions

Number

(MR) **1** $48 \times 52 = 2496$

Use this to help you work out the following calculations:

$24 \times 52 =$ _____ $48 \times$ _____ $= 1248$ $2496 \div 52 =$ _____   [3]

(PS) **2** The lowest common multiple of two numbers is 60 and their sum is 27

What are the numbers?   [2]

(FS) **3** Gemma is having a barbecue and wants to invite some friends.

Sausages come in packs of 6. Rolls come in packs of 8. She needs exactly the same number of sausages and rolls.

What is the minimum number of each pack she can buy?   [3]

Total Marks _____ / 8

---

1 Look at these expressions:

$45 = 5 \times 3^x$ $54 = 2 \times 3^y$

a) Find the values of x and y.   [2]

$45 \times 54 = 5 \times 2 \times 3^z$

b) Write down the value of z.   [1]

2 x is an integer. Explain why $25x^2$ is a square number.   [1]

(PS) **3** Two numbers have a sum of –5 and a product of 4

Write down the two numbers.   [1]

(MR) **4** Tom states that the sum of a square number and a cube number is always positive.

Is he right? Give an example to justify your answer.   [2]

Total Marks _____ / 7

Sequences

1 An expression for the nth term of the arithmetic sequence 6, 8, 10, 12,... is $2n + 4$

 a) Find the 20th term of this sequence. [1]

 b) Find the 100th term of this sequence. [1]

2 The odd numbers form an arithmetic sequence with a common difference of 2

 Find the nth term for the sequence of odd numbers. [2]

3 Lynne plants a new flower bush in her garden. Five of the buds have already flowered. Each week another three buds flower.

 a) How many buds will have flowered after three weeks? [1]

 b) How many weeks will it take for 32 buds to have flowered? [1]

 c) If n represents the number of weeks since Lynne planted her flower, write a rule to represent how many buds will flower after n weeks. [1]

Total Marks _____ / 7

(MR) 1 The nth term for a sequence of numbers is $4n^2 + 2$

 Wasim thinks the 10th term is 1602

 Cindy thinks the 10th term is 402

 Who is right?

 Explain your answer. [2]

2 Corinna visited the opticians to be fitted with some contact lenses. She is advised to wear them for three hours on the first day, and increase this by 20 minutes each day.

 After how many days will she be able to wear her contact lenses for 12 hours? [2]

Total Marks _____ / 4

Perimeter and Area

(PS) 1 The area of the rectangle shown is 48 cm²

Find the values of x and y. [2]

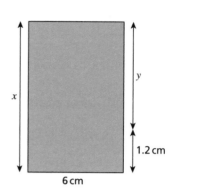

(FS) 2 Kelly is tiling a wall in her bathroom.

The wall is 4 m by 3 m. Each tile is 25 cm by 25 cm.

a) Work out how many tiles Kelly needs to buy for the wall. [3]

The tiles come in packs of 10 and each pack costs £15

b) Work out how much it will cost Kelly to tile the wall. [2]

c) How many tiles will she have left over? [1]

Total Marks _____ / 8

(PS) 1 The diagram shows a rhombus inside a rectangle.

The vertices of the rhombus are the midpoints of the sides of the rectangle.

Find the area of the rhombus. [3]

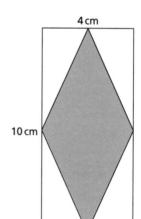

Total Marks _____ / 3

Statistics and Data

(MR) (1) Phil and Dave are good darts players.

They record their scores for a match. Their results are shown below.

Phil	64	70	80	100	57	100	41	56	30
Dave	36	180	21	180	10	5	23	25	140

a) Calculate the mean score for each player. [2]

b) Find the range of scores for each player. [2]

c) One of the two players can be picked to play in the next match.

Would you pick Phil or Dave? Explain your answer. [2]

Total Marks .... / 6

(1) The grouped frequency table below gives details of the weekly rainfall in a town in Surrey over a year.

Weekly rainfall in mm	Number of weeks
$0 \leqslant d < 10$	20
$10 \leqslant d < 20$	18
$20 \leqslant d < 40$	10
$40 \leqslant d < 60$	4

Estimate the mean weekly rainfall. [3]

Total Marks .... / 3

Decimals 1

You must be able to:

- Multiply and divide by 10, 100 and 1000
- Understand the powers of 10
- Understand standard form
- List decimals in size order.

Quick Recall Quiz

Multiplying and Dividing by 10, 100 and 1000

- Multiplying by 10 moves the digits one place to the left.
- Multiplying by 100 moves the digits two places to the left.
- Multiplying by 1000 moves the digits three places to the left.

$$1.67 \times 10 = 16.7 \qquad 1.67 \times 100 = 167 \qquad 1.67 \times 1000 = 1670$$

- Dividing by 10 moves the digits one place to the right.
- Dividing by 100 moves the digits two places to the right.
- Dividing by 1000 moves the digits three places to the right.

$$360.7 \div 10 = 36.07 \quad 360.7 \div 100 = 3.607 \quad 360.7 \div 1000 = 0.3607$$

- You can use one calculation to work out the answer to another.

Use $32 \times 24 = 768$ to work out:
a) 320×240 **b)** 3.2×2.4 **c)** 16×24

$\quad$ **a)** $320 \times 240 = 32 \times 10 \times 24 \times 10 = 768 \times 100 = 76\,800$

$\quad$ **b)** $3.2 \times 2.4 = 32 \div 10 \times 24 \div 10 = 768 \div 100 = 7.68$

$\quad$ **c)** $16 \times 24 = \frac{1}{2} \times 32 \times 24 = \frac{1}{2} \times 768 = 384$

- A decimal that does **not** recur is called a **terminating decimal**.

Powers of Ten and Standard Form

- A **power** or **index** tells us how many times a number should be multiplied by itself.
- A power is also called an **exponent**.

$$10^2 = 10 \times 10 = 100 \qquad\qquad 10^{-1} = \frac{1}{10}$$
$$10^3 = 10 \times 10 \times 10 = 1000$$
$$10^4 = 10 \times 10 \times 10 \times 10 = 10\,000 \qquad 10^{-2} = \frac{1}{10^2} = \frac{1}{100}$$

- **Standard form** or **standard index form** allows us to write very big and very small numbers more easily. Standard form uses powers of 10

> **Key Point**
>
> Multiplying by 10, 100 or 1000 makes the number bigger.

> **Key Point**
>
> Dividing by 10, 100 or 1000 makes the number smaller.

- A number **not** written in standard form is an **ordinary number**.
- Standard form is a number written in the form $A \times 10^n$, where $1 \leqslant A < 10$ and n is a positive or negative integer or zero.

2000 can be written as 2×1000, which is the same as 2×10^3

0.0007 can be written as $7 \div 10\,000$, which is 7×10^{-4}

4.56×10^5 can be written as $4.56 \times 100\,000$, so as an ordinary number is $456\,000$

7.62×10^{-3} can be written as $7.62 \times \dfrac{1}{1000}$ or $7.62 \div 1000$, so as an ordinary number is $0.007\,62$

Put these numbers in order of size. Start with the smallest.

$3.1 \times 10^3 \qquad 2.8 \times 10^5 \qquad 2.6 \times 10^4 \qquad 4.6 \times 10^3$

Starting with the numbers with the lowest powers of 10, as $3.1 < 4.6$, the order is:

$3.1 \times 10^3 \qquad 4.6 \times 10^3 \qquad 2.6 \times 10^4 \qquad 2.8 \times 10^5$

Ordering Decimals

- Place value can be used to compare decimal numbers.
- The digits after the decimal point are called tenths, hundredths, thousandths, and so on.

Put in ascending order: 12.071, 12.24, 12.905, 12.902, 12.061

Each number starts with 12. So compare the tenths, hundredths and thousandths.

	Tens	Units	.	Tenths	Hundredths	Thousandths
12.071	1	2	.	0	7	1
12.24	1	2	.	2	4	0
12.905	1	2	.	9	0	5
12.902	1	2	.	9	0	2
12.061	1	2	.	0	6	1

Key Point

Ascending order is smallest to biggest.

Descending order is biggest to smallest.

12.071 and 12.061 are the two smallest as they have no tenths.

12.24 is the next smallest with 2 tenths.

12.905 and 12.902 are the two biggest as they have 9 tenths.

First group by the number of tenths.

12.061 is smaller than 12.071 as it has only 6 hundredths compared to 7 hundredths.

Then order them within each group.

12.902 is smaller than 12.905, as although they both have the same hundredths, 12.902 has only 2 thousandths compared to 5 thousandths.

So, in ascending order: 12.061, 12.071, 12.24, 12.902, 12.905

Key Words

terminating decimal
power
index
exponent
standard form
ordinary number

Quick Test

1. a) Work out 23.56×10 b) Work out $56.781 \div 10$
2. Write down the value of 10^5
3. Put the numbers in ascending order:
   16.34, 16.713, 16.705, 16.309, 16.2

Decimals 2

You must be able to:

- Add and subtract decimal numbers
- Multiply and divide decimal numbers
- Use rounding to estimate calculations.

Adding and Subtracting Decimals

- Decimal numbers can be added and subtracted in the same way as whole numbers.

Calculate 23.764 + 12.987

	2	3	.	7	6	4
+	1	2	.	9	8	7
	3	6 ¹	.	7	¹5	¹1

So 23.764 + 12.987 = 36.751

Calculate 12.697 − 8.2

	⸒1	¹2	.	6	9	7
−	0	8	.	2	0	0
		4	.	4	9	7

So 12.697 − 8.2 = 4.497

Multiplying and Dividing Decimals

- Complete the calculation without the decimal points and replace the decimal point at the end.
- Count how many digits are after the decimal points in the question and this is how many digits are after the decimal point in the answer.

> **Key Point**
>
> When adding and subtracting, line up the numbers by matching the decimal point.

Calculate 45.3×3.7

Here using the column method gives $453 \times 37 = 16\ 761$

		4	5	3	
×			3	7	
	3	1 ³	7 ²	1	(7 × 453)
1	3 ¹	5	9	0	(30 × 354)
1	6	7 ¹	6	1	

There are two digits after the decimal point in the question.
So $45.3 \times 3.7 = 167.61$

- **Equivalent** fractions can be used when dividing decimals.

Calculate $4.45 \div 0.05$

$$4.45 \div 0.05 = \frac{4.45}{0.05} = \frac{445}{5}$$

$$5\overline{)4\,{}^4 4\,{}^4 5}$$

           8   9

So $4.45 \div 0.05 = 89$

Remember to multiply the numerator and denominator by the same amount.

> **Key Point**
>
> A division can be written as a fraction. Equivalent fractions are equal.

Rounding and Estimating

- Numbers can be **rounded** using **decimal places** (d.p.) or **significant figures** (s.f.).

> Round 56.76 to 1 decimal place.
>
> 56.7|6
>
> 7 is the first decimal place and the digit after it is more than 5 so round 7 up to 8. 56.76 to 1 decimal place is 56.8

> Round 0.00764 to 2 significant figures.
>
> 0.0076|4
>
> 6 is the second significant figure and the digit after it is less than 5, so the 6 stays as a 6
>
> 0.00764 rounded to 2 significant figures is 0.0076

 Remember that the first significant figure is the first non-zero digit on the left.

- When **estimating** a calculation, round all the numbers to 1 s.f.

> Estimate 26 751 × 64
>
> An estimate for 26 751 × 64 is 30 000 × 60 = 1 800 000

- There is always a resulting error from approximations and estimates.

> 26 751 × 64 = 1 712 064
>
> An estimate for 26 751 × 64 = 1 800 000
>
> The resulting error is 1 800 000 − 1 712 064 = 87 936

Key Point

The first significant figure is the first non-zero digit.

- There is always an error to consider when a number is rounded.
- This error can be expressed using an inequality.

> A number has been rounded to 7.6 to 1 decimal place. The diagram below shows the range of possible values the number could be.
>
> 7.5   7.6   7.7
>
>   7.55   7.65
>
> The number could be anywhere between 7.55 and 7.65, therefore the rounding error can be expressed as
> −0.05 ⩽ error < 0.05

Key Words

equivalent
rounding
decimal places
significant figures
estimate

Quick Test

1. Work out 45.671 + 3.82
2. Work out 34.321 − 17.11
3. Work out 65.2 ÷ 0.4
4. Estimate 3457 × 46

Algebra 1

You must be able to:

- Understand the vocabulary associated with algebra
- Know the difference between an equation and expression
- Collect like terms in an expression
- Write products as algebraic expressions
- Substitute numerical values into formulae and expressions.

Vocabulary in Algebra

- The numbers used in algebra are called **constants**.
- The letters used in algebra are called **variables**.
- A **term** is part of an **expression**, **equation** or **formula**.
- An expression is a collection of terms.
- An equation has an equals sign.
- A formula is a rule which links a variable to one or more other variables.

$2w + 3y + 6$ is an expression	$2w = 6$ is an equation	$P = 2w + 2l$ is a formula
$2w$ is a term	6 is a constant	w is a variable

- The number in front of a variable in a term is called the **coefficient**. For example, in the term $2w$ the coefficient is 2

Collecting Like Terms

- **Like** terms are terms with the same variables.
- To **simplify** an expression, like terms are collected.

Simplify $5x + 8y + 3x - y$

Collect the like terms:

$5x + 3x + 8y - y$

The x terms can be simplified: $5x + 3x = 8x$

The y terms can be simplified: $8y - y = 7y$

So $5x + 8y + 3x - y$ can be simplified to $8x + 7y$

⟵ In this example there are x terms and y terms.

 Key Point

Remember that terms have a + or − sign between them and each sign belongs to the term on its right.

Simplify $2x^2 + 6y - x^2 + 4y - 6$

Collect the like terms: $2x^2 - x^2 + 6y + 4y - 6$

The x^2 terms can be simplified: $2x^2 - x^2 = x^2$

The y terms can be simplified: $6y + 4y = 10y$

There is only one constant term: $- 6$

So $2x^2 + 6y - x^2 + 4y - 6$ can be simplified to $x^2 + 10y - 6$

⟵ In this example, there are x^2 terms, y terms and constant terms.

Simplify $\frac{2}{3}x + y - \frac{1}{3}x + \frac{3}{4}y$

Collect the like terms: $\frac{2}{3}x - \frac{1}{3}x + y + \frac{3}{4}y$

The x terms can be simplified: $\frac{2}{3}x - \frac{1}{3}x = \frac{1}{3}x$

The y terms can be simplified: $y + \frac{3}{4}y = \frac{7}{4}y$

So $\frac{2}{3}x + y - \frac{1}{3}x + \frac{3}{4}y$ can be simplified to $\frac{1}{3}x + \frac{7}{4}y$

> In this example, some of the coefficients are fractions.

Expressions with Products

- The **product** of a and b is the same as $a \times b$ or ab
- Expressions with products are usually written without the $\times$ sign.

$2 \times a = 2a$

$a \times b = ab$

$a \times a = a^2$

$a \times a \times b = a^2b$

$a \times a \times a = a^3$

$a \div (b \times c) = \dfrac{a}{bc}$

> **Key Point**
>
> There are many scientific problems which involve substituting into formulae. Always follow BIDMAS.
>
> To find the perimeter, area and volume of shapes, we substitute into a formula.

Substitution

- Variables in formulae can be written in shorthand by representing them with a letter.
- Some commonly used scientific formulae are:

$\text{speed} = \dfrac{\text{distance}}{\text{time}}$ in shorthand $s = \dfrac{d}{t}$

$\text{density} = \dfrac{\text{mass}}{\text{volume}}$ in shorthand $d = \dfrac{m}{v}$

- Substitution involves replacing the letters in a given formula or expression with numbers.

Find the value of the expression $2a^3 + b$ when $a = 3$ and $b = 5$

Replace the letters with the given numbers.

$2a^3 + b = 2 \times 3^3 + 5 = 59$

$\frac{3}{4} = 0.75$

Emma took $\frac{3}{4}$ of an hour to travel 30 miles.

Calculate her average speed in miles per hour.

$s = \dfrac{d}{t}$

$s = \dfrac{30}{0.75} = 40\,\text{mph}$

> **Key Words**
>
> constant
> variable
> term
> expression
> equation
> formula
> coefficient
> like terms
> simplify
> product

> **Quick Test**
>
> 1. Simplify $4x + 7y + 3x - 2y + 6$
> 2. Simplify $c \times c \times d \times d$
> 3. Find the value of $4x + 2y^2$ when $x = 2$ and $y = 3$

Algebra 2

You must be able to:

- Understand the commutative, associative and distributive laws
- Multiply a single term over a bracket
- Carry out binomial expansion
- Factorise linear expressions
- Change the subject of a formula.

Fundamental Laws

- The **commutative law** says that $a + b = b + a$, e.g. $2 + 3 = 3 + 2$
- The **associative law** says that $(a + b) + c = a + (b + c)$, e.g. $(2 + 3) + 4 = 2 + (3 + 4)$
- The **distributive law** says that $a(b + c) = ab + ac$, e.g. $2(3 + 4) = 2 \times 3 + 2 \times 4$

Work out the answers to check these are true.

Key Point

A reminder of the rules when multiplying and dividing negative numbers:

	+	−
+	+	−
−	−	+

Expanding Brackets

- **Expanding** the brackets involves removing the brackets by **multiplying every term** inside the bracket by the number or term on the outside.

Expand and simplify $4(x + y) - 2(2x - 3y)$

×	4
x	$4x$
y	$4y$

$= 4x + 4y$

×	2
$2x$	$4x$
$-3y$	$-6y$

$= 4x - 6y$

Then collect like terms: $(4x + 4y) - (4x - 6y) = 10y$

Expand $2x^3\left(5x^2 - 6y^2\right)$

×	$2x^3$
$5x^2$	$10x^5$
$-6y^2$	$-12x^3y^2$

$2x^3\left(5x^2 - 6y^2\right) = 10x^5 - 12x^3y^2$

↑ This is the distributive law.

Binomial Expansion

- A **binomial** is an expression that contains two terms, e.g. $3y - 1$
- The product of two (or more) binomials is when they are multiplied together, e.g. $(x + 4)(3x - 1)$
- To expand (or multiply out) the brackets, every term in the first bracket is multiplied by every term in the second bracket.

Expand and simplify $(3y - 2)(2y + 7)$

×	$3y$	-2
$2y$	$6y^2$	$-4y$
$+7$	$+21y$	-14

$(3y - 2)(2y + 7) = 6y^2 - 4y + 21y - 14$
$= 6y^2 + 17y - 14$

← Simplify by collecting like terms.

Expand and simplify $(x + 3)(x + 5)(x + 1)$

×	x^2	$8x$	$+15$
x	x^3	$8x^2$	$+15x$
$+1$	x^2	$8x$	$+15$

$(x + 3)(x + 5) = x^2 + 8x + 15$
So $(x + 3)(x + 5)(x + 1) = (x^2 + 8x + 15)(x + 1)$
$= x^3 + 8x^2 + 15x + x^2 + 8x + 15$
$= x^3 + 9x^2 + 23x + 15$

← Simplify by collecting like terms.

Factorising

- Factorising (or factorisation) is the reverse of expanding brackets.
- To factorise, look for **common factors**.

Factorise $6x + 9$

×	3
	6x
	+9

×	3
2x	6x
+3	+9

3 is a common factor of 6 and 9 so take the 3 to the outside of the bracket. To find what is inside the bracket, fill in the blanks in the table.

So $6x + 9 = 3(2x + 3)$

Factorise $6x^3 + 2x^2$

2 and x^2 are the common factors:

So $6x^3 + 2x^2 = 2x^2(3x + 1)$

×	2x²
3x	6x³
+1	2x²

Factorise $\frac{7}{9}xy + \frac{8}{9}y^2$

Both $\frac{1}{9}$ and y are common factors.

×	$\frac{1}{9}y$
	$\frac{7}{9}xy$
	$\frac{8}{9}y^2$

×	$\frac{1}{9}y$
7x	$\frac{7}{9}xy$
8y	$\frac{8}{9}y^2$

So $\frac{7}{9}xy + \frac{8}{9}y^2 = \frac{1}{9}y(7x + 8y)$

Changing the Subject of a Formula

- The **subject** of a formula appears once on the left-hand side.
- To change the subject, a formula must be rearranged using **inverse** operations. Write the rearranged formula with the new subject on the left-hand side.

Make x the subject of $y = 3x + 1$

$y - 1 = 3x$ ← Subtracting 1 from both sides.

$\frac{y-1}{3} = x$ or $x = \frac{y-1}{3}$ ← Dividing both sides by 3

Make r the subject of $A = \pi r^2$ ← This is the formula for the area of a circle with radius r.

$\frac{A}{\pi} = r^2$ ← Dividing both sides by π

$\sqrt{\frac{A}{\pi}} = r$ or $r = \sqrt{\frac{A}{\pi}}$ ← Taking the square root of both sides.

← This is the formula for the radius of a circle with area A.

Make x the subject of $y = \frac{x-5}{x+2}$

$(x + 2)y = x + 5$ ← Multiplying both sides by $x + 2$

$xy + 2y = x + 5$ ← Expanding the bracket.

$xy - x = 5 - 2y$ ← Collecting terms in x to the left and other terms to the right.

$x(y - 1) = 5 - 2y$

$x = \frac{5 - 2y}{y - 1}$ ← Factorising.

← Dividing both sides by $y - 1$

Quick Test

1. Expand $4(2x - 1)$
2. Expand and simplify $2(2x - y) - 2(x + 6y)$
3. Factorise $5x - 25$
4. Factorise completely $2x^5 - 4x^3$

Review Questions

Perimeter and Area

 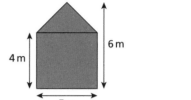

1 Frances wants to paint the side of her house.
The diagram represents the side of Frances' house.

   a) Find the area that Frances needs to paint. [3]

Each tin of paint covers $7\,m^2$

   b) Work out the number of tins that Frances needs to buy. [1]

Each tin costs £12

   c) Work out how much it will cost Frances to paint the side of her house. [2]

2 Yoon buys a new bicycle and uses it to cycle 8000 m to work every day. His wheel is a circle with a diameter of 50 cm.

Work out how many times his wheel makes a full rotation during his journey. [2]

Total Marks .... / 8

---

 1   **a)** Which of the two sectors below has the bigger area? Show working to justify your answer. [3]

Sector A   Sector B

Radius 3 cm   Radius 5 cm
$\frac{1}{5}$ of a circle $\frac{1}{9}$ of a circle

   b) The perimeter is made up of two straight edges and the arc length.

Which of the two sectors above has the bigger perimeter?
Show working to justify your answer. [3]

Total Marks .... / 6

Statistics and Data

(PS) **1** A midwife asked 60 of her patients if they wanted a home, hospital or water birth.

16 of her patients were teenagers.

10 of the teenagers wanted a hospital birth.

18 of the non-teenagers wanted a home birth.

4 of the 10 patients who wanted a water birth were teenagers.

How many patients wanted a hospital birth? [3]

(MR) **2** The data below shows the number of people attending the first six home matches for Sandex United football club:

| 1240 | 1354 | 1306 | 14 808 | 1378 | 1430 |

You want to calculate the average attendance.

Would you find the mean, median or mode? Give a reason for your answer. [1]

Total Marks / 4

(MR) **1** Sophie carries out a survey of shoe size in her class. There are 28 pupils in her class. She calculates the mean to be 7.5

She thinks this is a little high, given her data, and decides to check her calculations. She realises she has used the value 40 instead of 4

Calculate the correct mean. [3]

(MR) **2** Isabella is organising a charity netball event. She is planning on selling T-shirts to help raise extra money. She is trying to decide how many of each size T-shirt to order, so she asks all the people in her class their T-shirt size.

Should Isabella use the mean, median or mode as the average size in this case? Give a reason to justify your answer. [2]

Total Marks / 5

Practice Questions

Decimals

1. Work out the following: 📱

   a) $34.542 + 23.29$ [1]

   b) $65.21 - 43.23$ [1]

   c) 21.81×3.4 [1]

   d) $43.2 \div 0.2$ [1]

2. Put these decimals in order of size. Start with the smallest.

   17.84, 17.203, 17.09, 16.94, 16.061 [2]

   Total Marks _____ / 6

1. Join the pairs of cards that multiply to make 4

   | 0.02 | | 0.08 |

   | 50 | | 0.2 |

   | 8 | | 200 |

   | 20 | | 0.5 |

   [2]

2. a) Write 6 890 000 in standard form. [1]

   b) Write 0.008 766 in standard form. [1]

   c) Which is larger, 5.989×10^4 or 59 890 000?
   Justify your answer. [1]

   Total Marks _____ / 5

Algebra

(MR) **1** Chanda thinks the perimeter of the rectangle is $2x + 2y$

Lawrence thinks the perimeter of the rectangle is $2(x + y)$

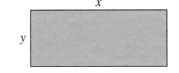

Who is right? Chanda, Lawrence or both of them?
Explain your answer. [2]

(FS) **2** The cost (in £) of hiring a car for the day and driving it y miles is shown by this formula:

$$C = 75 + 0.4y$$

Work out how much it would cost to hire the car for a day and travel 120 miles. [3]

3 Expand and simplify $2(4x + 1) - 5(x - 1)$ [2]

4 Factorise completely $3abc + 6a$ [2]

(PS) **5** $ab = 36 \qquad a = 4$

Find the value of a^2b [2]

6 Copy and complete the algebra grid. Each brick is made by the sum of the two bricks underneath.

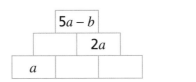 [2]

Total Marks .... / 13

1 Expand the brackets:

$(x + 4)(x - 2)$ [2]

2 Complete the following factorisations:

$x^2 + 3x + 2 = (x + 1)(\underline{\hspace{0.5cm}} + \underline{\hspace{0.5cm}})$

$x^2 + 5x - 6 = (x + 6)(\underline{\hspace{0.5cm}} - \underline{\hspace{0.5cm}})$ [2]

(MR) **3** Kathleen states that for all numbers $(x + y)^2 = x^2 + y^2$

Show that Kathleen is wrong. [2]

Total Marks .... / 6

3D Shapes: Volume and Surface Area 1

You must be able to:

- Name and draw 3D shapes
- Draw the net of a cuboid and other 3D shapes
- Calculate the surface area and volume of a cuboid.

Quick Recall Quiz

Naming and Drawing 3D Shapes

- A 3D shape can be described using the number of **faces**, **vertices** and **edges** it has.
- A **prism** is a solid shape with a uniform **cross-section**. The cross-section of a prism can be any polygon.
- Cubes and cuboids are prisms.

Shape	Name	Edges	Vertices	Faces
	Cube	12	8	6
	Cuboid	12	8	6
	Triangular prism	9	6	5
	Square-based pyramid	8	5	5
	Cylinder	2	0	3
	Pentagonal prism	15	10	7
	Cone	1	1	2
	Sphere	0	0	1

> ### Key Point
>
> A face is a sur'face', for example a flat side of a cube.
>
> A vertex is where edges meet, for example the corner of a cube.
>
> An edge is where two faces join.

Nets for 3D Shapes

- To create the **net** of a cuboid, imagine it is a box you are unfolding to lay out flat.

- Nets for a triangular prism and a square-based pyramid look like this:

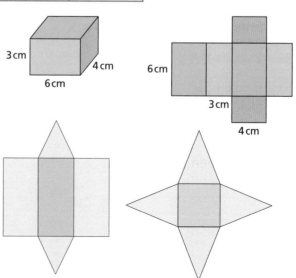

Surface Area of a Cube or Cuboid

- The **surface area** of a cube or cuboid is the sum of the areas of all six faces. The units for area are cm^2, m^2, etc. On a cube all the faces have the same area.

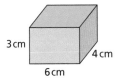

To calculate the surface area of the cuboid shown right, you can use the net to help you. Work out the area of each rectangle by multiplying the base by its height:

Green rectangle: 6 cm × 4 cm = 24 cm^2
There are two of them, so 24 cm^2 × 2 = 48 cm^2

Blue rectangle: 3 cm × 6 cm = 18 cm^2
There are two of them, so 18 cm^2 × 2 = 36 cm^2

Pink rectangle: 4 cm × 3 cm = 12 cm^2
There are two of them, so 12 cm^2 × 2 = 24 cm^2

Sum of all six areas is 48 + 36 + 24 = 108 cm^2

Volume of a Cuboid

- **Volume** is the space contained inside a 3D shape.

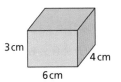

To calculate the volume of the cuboid shown right, you need to first work out the area of the front rectangle:

3 cm × 6 cm = 18 cm^2

Next multiply this area by the depth of the cuboid, 4 cm.

18 cm^2 × 4 cm = 72 cm^3

⟵ The units are cm^3 this time.

- You can use the formula Volume of a cuboid (V) = Length (l) × width (w) × height (h) or $V = lwh$

Find the volume of a cuboid measuring 6 cm by 3 cm by 2 cm.

$V = 6 \times 3 \times 2$
$\quad = 36$ cm^3

Key Words

face
vertex
edge
prism
cross-section
net
surface area
volume

Quick Test

Work out the volume and the surface area of these cuboids.

1.
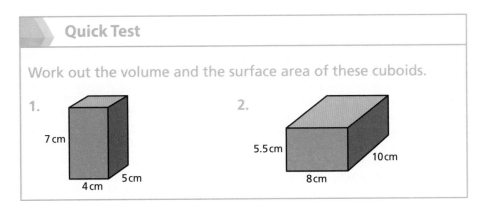

2.

3D Shapes: Volume and Surface Area 2

You must be able to:

- Calculate the volume and surface area of a cylinder
- Calculate the volume and surface area of a prism
- Calculate the volume of composite shapes.

Volume and Surface Area of a Cylinder

- You work out the volume of a **cylinder** the same way as the volume of a cuboid. First work out the area of the circle and then multiply it by the height of the cylinder.

Calculate the volume of this cylinder.

Volume = $(\pi \times 5 \times 5) \times 3$

       = $235.62\,\text{cm}^3$ (2 d.p.)

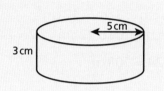

Calculate the volume of this cylinder.

The cylinder has a diameter of 12 cm.

This means the radius is 6 cm.

Volume = $(\pi \times 6 \times 6) \times 5$

       = $565.49\,\text{cm}^3$ (2 d.p.)

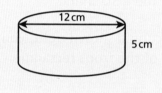

- To calculate the surface area of a cylinder, first draw the net (imagine cutting a can open).

Calculate the surface area of this cylinder.

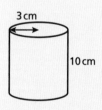
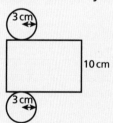

The circumference of the circular end is the same as the length of the rectangle.

Calculate the area of the circular ends $(\pi \times 3 \times 3) \times 2 = 56.548...$

The area of the rectangle is the circumference of the circle multiplied by the height of the cylinder, 10 cm.

$(\pi \times 6) \times 10 = 188.495...$

Add the areas together: $56.548... + 188.495... = 245.04\,\text{cm}^2$ (2 d.p.)

- You can use the formula
Curved surface area of a cylinder $C = 2\pi r h$ or $C = \pi d h$

Key Point

Diameter is the full width of a circle that goes through the centre.

Radius is half of the diameter.

Area of a circle = $\pi \times \text{radius}^2$

Volume units are shown by a ^{3}, for example cm^3

Key Point

Circumference of a circle is the perimeter.

Because there are two circles.

Circumference = $\pi \times$ diameter
Diameter is double the radius.

Volume and Surface Area of a Prism

- Volume of prism = Area of cross-section × length or $V = Al$
- The surface area of a prism is the sum of the areas of all the faces.

Work out the volume and surface area of the triangular prism.

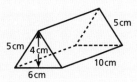

Area of cross-section is $\frac{1}{2} \times 6 \times 4 = 12\,cm^2$

Volume of prism is $12 \times 10 = 120\,cm^3$

Surface area =

2 × area of cross-section + area of the three rectangles

$= 12 + 12 + (10 \times 6) + (10 \times 5) + (10 \times 5)$

$= 12 + 12 + 60 + 50 + 50$

$= 184\,cm^2$

Volume of Composite Shapes

- **Composite** means the shape has been 'built' from more than one shape.

This shape is built from two cuboids.

Calculate the volume of the two separate cuboids and add the volumes together.

$(6\,cm - 2\,cm) \times 5 \times 3 = 60\,cm^3$

$(3\,cm + 4\,cm) \times 2 \times 5 = 70\,cm^3$

$60 + 70 = 130\,cm^3$

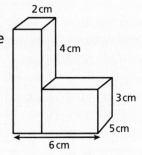

> **Key Point**
>
> Use the shape's dimensions to work out missing lengths.

Quick Test

1. Work out the volume and the surface area of these shapes to the nearest whole unit.

   a)

   b)

2. Work out the volume of this composite shape.

> **Key Word**
>
> cylinder

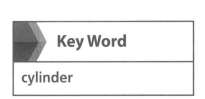

Interpreting Data 1

Statistics

You must be able to:

- Create a simple pie chart from a set of data
- Create and interpret pictograms
- Use frequency tables and draw frequency diagrams
- Make comparisons and contrasts between data.

Pie Charts

- **Pie charts** are often shown with **percentages** or **angles** indicating sector size – this and the visual representation helps to interpret the **data**.

36 students were asked the following question:
Which is your favourite flavour of crisps?
To work out the angle for
each person: 360 ÷ 36 = 10°

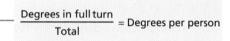
$$\frac{\text{Degrees in full turn}}{\text{Total}} = \text{Degrees per person}$$

Flavour of crisps	No. of students	Degrees
Salt and Vinegar ■	8	8 × 10 = 80
Cheese and Onion ■	10	10 × 10 = 100
Ready Salted ■	12	12 × 10 = 120
Prawn Cocktail ■	4	4 × 10 = 40
Other ■	2	2 × 10 = 20

> **Key Point**
>
> Always align your protractor's zero line with your starting line.
>
> Count up from zero to measure your angle.
>
> Don't forget to label your chart.

Pictograms

- Data is represented by a picture or symbol in a **pictogram**.

The pictogram shows how many pizzas were delivered by Ben in one week. Key = 8 pizzas

Day	Mon	Tue	Wed	Thu	Fri	Sat	Sun
Pizza deliveries	🍕🍕 🍕	🍕	🍕◖	🍕	🍕🍕 🍕🍕	🍕🍕	

How many pizzas did Ben deliver on Friday? 8 × 4 = 32

On Sunday Ben delivered 20 pizzas. Complete the pictogram.

20 ÷ 8 = 2.5

Frequency Tables and Frequency Diagrams

- **Frequency tables** can be used to collect data into groups.
- **Frequency diagrams** (or frequency charts) illustrate the information.

The frequency table shows the heights of 100 students.

Height (h cm)	Frequency
$70 < h \leqslant 80$	3
$80 < h \leqslant 90$	4
$90 < h \leqslant 100$	12
$100 < h \leqslant 110$	24
$110 < h \leqslant 120$	30
$120 < h \leqslant 130$	22
$130 < h \leqslant 140$	3
$140 < h \leqslant 150$	2

Using the span of each height category, plot each group as a block using the frequency **axis**.

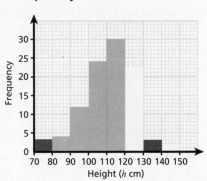

Data Comparison

- You can make comparisons using frequency diagrams.

These graphs show how much time four students spend on their mobile phones in a week. Compare the two graphs.

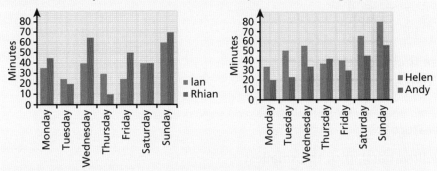

Apart from Thursday, Helen uses her phone more than Andy. In Ian's and Rhian's data, Rhian uses her phone more overall. Comparing both sets of data, we can say the girls use their phones more than the boys.

Key Point

Think about what is similar about the data and what is different. Look for any patterns.

Quick Test

1. If 18 people were asked a question and you were to create a pie chart to present your data, what angle would represent one person?
2. Using the graph above right, on which day did both Helen and Andy use their mobiles the most?
3. The following week, Andy had a mean use of 40 minutes per day. Is this greater than Helen's use the previous week?

Key Words

pie chart
percentage
angle
data
pictogram
frequency table
frequency diagram
axis

Quick Recall Quiz

Interpreting Data 2

You must be able to:

- Interpret different graphs and diagrams
- Draw a scatter graph and understand correlation
- Understand the use of statistical investigations.

Interpreting Graphs and Diagrams

- You can interpret the information in graphs and diagrams.

What does this graph show?

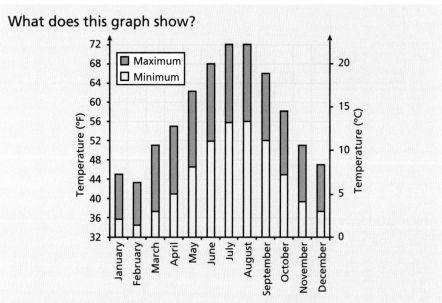

Using the labels and key, you can see that it gives the temperature and months of a year, plus a maximum and minimum temperature.

Which month has the lowest temperature?

Using the minimum temperature (yellow bars), pick the smallest bar, in this case, February.

Look at this pie chart. What is the most likely way the team scores a goal?

The largest sector of the pie chart represents scoring a goal in free play.

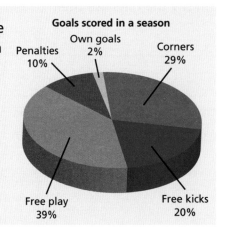

Goals scored in a season

Penalties 10%
Own goals 2%
Corners 29%
Free kicks 20%
Free play 39%

Drawing a Scatter Graph

- A **scatter graph** is used to investigate a relationship between two variables. Data for two variables is called **bivariate data**.
- If a linear relationship exists, a **line of best fit** can be drawn through the points on a scatter graph.

> Here is a plot of the sales of ice cream against the amount of sun per day for 12 days. The scatter graph shows that when the weather is sunnier (and hotter), more ice creams are sold.
>
>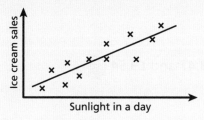

- You can describe the **correlation** and use the line of best fit to estimate data values. The graph above shows a positive correlation between ice cream sales and sunlight in a day.

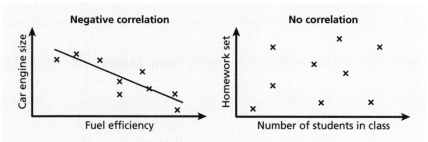

Key Point

Plot each data pair as a coordinate. A line of best fit doesn't have to start at zero.

This scatter graph is showing bivariate data.

Statistical Investigations

- Statistical investigations use **surveys** and experiments to test statements and theories to see whether they might be true or false. These statements are called **hypotheses**.

> Jia needs to throw a 1 on a dice. She rolls it 25 times and still hasn't rolled a 1. What hypothesis might Jia make?
>
> Hypothesis: The dice is biased.
>
> You could test this hypothesis by rolling the dice a large number of times, determining whether it favours a certain number or not.

Key Point

Surveys should be made on a large random sample, never just a limited few.

Quick Test

1. Look at the pie chart on page 48. What is the least likely way that the team scored a goal?
2. What correlation might you expect when comparing umbrella sales and rainfall?
3. Richard hasn't spun a yellow in 20 spins; his spinner has five different colours on it. What hypothesis might Richard suggest?

Key Words

scatter graph
bivariate data
line of best fit
correlation
survey
hypothesis

Review Questions

Decimals

1 Put the following numbers in order from smallest to biggest:

7.765, 7.675, 6.765, 7.756, 6.776 [2]

2 Tomas buys three books which cost £2.98, £3.47 and £9.54

a) How much did the books cost in total? [2]

b) How much change should he get from a £20 note? [1]

Total Marks _____ / 5

FS 1 Louisa wants to buy some stationery. Here is a list of what she wants to buy:

Pencil 65p

Ruler £1.20

Pack of pens £4.99

Folder 98p

Eraser 48p

a) Find an estimate for the cost of her stationery. You should round all costs to 1 significant figure. [1]

b) Calculate the exact cost of her purchases. [1]

c) Find the percentage error in the estimate. [2]

Total Marks _____ / 4

Algebra

(MR) (1) This rectangle has dimensions $a \times b$

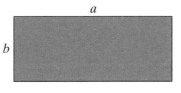

a) Write a simplified expression for the area of this rectangle. [1]

b) Write a simplified expression for the perimeter of this rectangle. [1]

c) Another rectangle has area $15a^2$ and perimeter $16a$.
   What are the dimensions of this rectangle? [1]

(2) $y = \dfrac{20}{\pm\sqrt{x+10}}$

When $x = 90$ there are two possible values of y. Write down both values. [2]

(3) Factorise the following expression completely: $8ut^2 - 4ut + 20t$ [2]

(4) Kieran states 'If n is a positive integer then $4n$ will always be even'. Is Kieran correct?
Explain your answer. [1]

Total Marks _____ / 8

(PS) (1) a) Expand the expression $(x + y)(x - y)$ [1]

b) Use the expression $(x + y)(x - y)$ to find the answer to $201^2 - 199^2$ [2]

(MR) (2) Jack is buying a new desk for his bedroom. He chooses a desk but wants to check it will
fit in the alcove in his bedroom. He measured the length of the desk to be 1.9 m to the
nearest centimetre. He measured his alcove to be 200 cm to the nearest 10 cm.

Desk
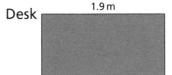
1.9 m

Alcove
200 cm

Will his desk definitely fit in the alcove? Justify your answer. [3]

Total Marks _____ / 6

Practice Questions

3D Shapes: Volume and Surface Area

1 Calculate the surface area and volume of this cuboid. 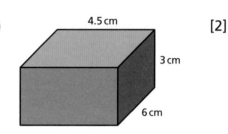 [2]

(MR) **2** Find the height of a cylinder with a radius of 5 cm and volume of 942 cm³
Give your answer to 1 decimal place. [2]

(MR) **3** Find the height of a cylinder with a radius of 7 cm and volume of 1385 cm³
Give your answer to 2 decimal places. [2]

Total Marks .... / 6

(MR) **1** Calculate the surface area and volume of this triangular prism.  [2]

(MR) **2** Find the volume of these shapes.

a) [6]

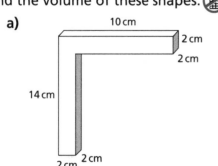

b)

Total Marks .... / 8

Interpreting Data

1. What two things might you plot against one another to show a negative correlation? [2]

2. Design a question with response box options to determine whether people shop more over the Christmas period than at other times of the year. [4]

(MR) 3. Name four different types of statistical graphs or charts.

Which one would you use to plot the information collected from asking 40 students 'How long do you spend doing homework in a week?' Explain your choice. [4]

(MR) 4. One hundred students take two maths tests. The results are shown on the graph.

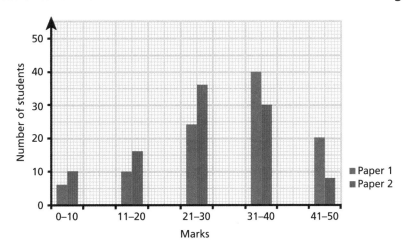

Which test did the students find more difficult? Give a reason for your answer. [2]

Total Marks _____ / 12

1. Zarifa rolls an eight-sided dice more than 30 times and never rolls an 8

What hypothesis might Zarifa suggest? How could she test her hypothesis? [3]

Total Marks _____ / 3

Fractions 1

You must be able to:

- Find equivalent fractions
- Order fractions
- Add and subtract fractions.

Equivalent Fractions

- Equivalent fractions are fractions that are equal despite the **denominators** being different.

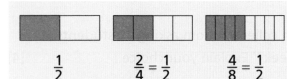

$$\frac{1}{2} \qquad \frac{2}{4} = \frac{1}{2} \qquad \frac{4}{8} = \frac{1}{2}$$

- You can create an equivalent fraction by keeping the ratio between the **numerator** and denominator the same.
- You do this by multiplying or dividing both the numerator and denominator by the same number.
- Creating equivalent fractions is very useful when you want to compare or evaluate different fractions.

Ordering Fractions

- You can use equivalent fractions to compare the size of fractions.

Which is the larger fraction, $\frac{2}{5}$ or $\frac{3}{7}$?

$\frac{2}{5}$ 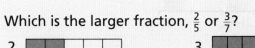 $\frac{3}{7}$

To compare these fractions, you need to find a common denominator – a number that appears in both the 5 and 7 times tables.

$5 \times 7 = 35$

So $\frac{2}{5}$ becomes $\frac{14}{35}$ and $\frac{3}{7}$ becomes $\frac{15}{35}$

Now the denominators are equal, you can compare the two fractions more easily and you can see that $\frac{15}{35} = \frac{3}{7}$ is larger.
So $\frac{3}{7} > \frac{2}{5}$

> ### Key Point
>
> The numerator is the top part of a fraction.
>
> The denominator is the bottom part of a fraction.

> ### Key Point
>
> A common denominator is a number that shares a relationship with both fractions' denominators.
>
> For example, for 5 and 3 this would be 15, 30, 45, 60, …

Adding and Subtracting Fractions

- Adding fractions with the same denominator is straightforward. The numerators are collected together.

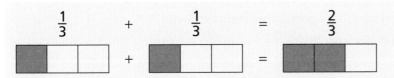

Notice the size of the 'piece', the denominator, remains the same in both the question and the answer.

- When subtracting fractions with the same denominator, simply subtract one numerator from the other.

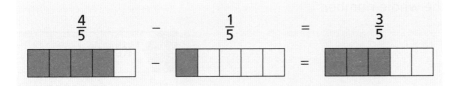

Key Point

The size of the 'piece' (the denominator) has to be the same to perform either addition or subtraction.

- When you have fractions with different denominators, first find equivalent fractions with a common denominator.

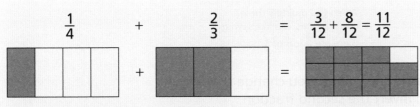

Here the common denominator is 12, as it is the smallest number that appears in both the 3 and 4 times tables.

This means that, for the first fraction, you have to multiply both the numerator and denominator by 3 and for the second fraction multiply them by 4

Now the fractions are of the same size 'pieces', you can add the numerators as before.

Key Point

It is essential to find equivalent fractions so both fractions have the same denominator.

Quick Test

1. Find three equivalent fractions for $\frac{2}{3}$

2. Work out $\frac{2}{7} + \frac{6}{11}$

3. Work out $\frac{7}{9} - \frac{3}{8}$

4. Work out $\frac{7}{13} - \frac{1}{4}$

5. Work out $\frac{14}{25} + \frac{3}{5} - \frac{7}{20}$

Key Words

denominator
numerator

Fractions 2

You must be able to:

- Multiply and divide fractions
- Understand mixed numbers and improper fractions
- Add, subtract, multiply and divide using mixed numbers.

Multiplying and Dividing Fractions

- Multiplying fractions by whole numbers is not very different from multiplying whole numbers.
- The numerator is multiplied by the whole number.

$$\frac{2}{7} \times 3 = \frac{2}{7} + \frac{2}{7} + \frac{2}{7} = \frac{6}{7}$$

> **Key Point**
>
> No common denominator is needed for multiplying or dividing fractions.

- When multiplying fractions, you multiply the numerators and then multiply the denominators.

$$\frac{3}{5} \times \frac{2}{7} = \frac{3 \times 2}{5 \times 7} = \frac{6}{35}$$

> When you multiply these two fractions, it is like saying there are $\frac{3}{5}$ lots of $\frac{2}{7}$

- When dividing fractions, use **inverse** operations. You change the operation to a multiplication and invert the second fraction.

$$\frac{2}{5} \div \frac{1}{2} = \frac{2 \times 2}{5 \times 1} = \frac{4}{5}$$

> $\frac{2}{5}$ divided into halves gives twice as many pieces.

Mixed Numbers and Improper Fractions

- A **mixed number** is where there is both a whole number part and a fraction, for example $1\frac{1}{3}$
- An **improper fraction** is where the numerator is bigger than the denominator, for example $\frac{4}{3}$

Change the improper fraction $\frac{14}{3}$ to a mixed number.

$$\frac{14}{3} = 4\frac{2}{3}$$

> How many 3s are there in 14?

> The remainder is left as a fraction.

Change the mixed number $5\frac{3}{4}$ to an improper fraction.

$$5\frac{3}{4} = \frac{(5 \times 4) + 3}{4} = \frac{23}{4}$$

> Multiply the whole number by the denominator. 5 units is 20 quarters.

> Add the $\frac{3}{4}$ to make 23 quarters.

Adding and Subtracting Mixed Numbers

- To add or subtract with mixed numbers:
  Convert any mixed numbers to improper fractions →
  Convert to equivalent fractions with a common denominator →
  Add (or Subtract) →
  Convert back to a mixed number

$$2\frac{4}{9} + 3\frac{1}{4} = \frac{22}{9} + \frac{13}{4} = \frac{22 \times 4}{36} + \frac{13 \times 9}{36} = \frac{205}{36} = 5\frac{25}{36}$$

$$4\frac{4}{7} - 1\frac{1}{4} = \frac{32}{7} - \frac{5}{4} = \frac{32 \times 4}{28} - \frac{5 \times 7}{28} = \frac{93}{28} = 3\frac{9}{28}$$

> **Key Point**
>
> To add, subtract, multiply or divide with mixed numbers, you first need to convert them to improper fractions.

Multiplying and Dividing Mixed Numbers

- To multiply mixed numbers, there are four steps:
  Convert any mixed numbers to improper fractions →
  Write down the multiplication and simplify the fractions by cancelling by any common factors in the numerators and denominators, if possible →
  Multiply the numerators to obtain the numerator of the answer and multiply the denominators to obtain the denominator of the answer →
  If the answer is improper, convert to a mixed number
- To divide mixed numbers:
  Convert any mixed numbers to improper fractions →
  Convert the division calculation into a multiplication calculation → **Carry out the multiplication**

$$2\frac{1}{4} \times 1\frac{3}{5} = \frac{9}{4} \times \frac{8}{5}$$
$$= \frac{9 \times 8}{4 \times 5} = \frac{9 \times \cancel{8}^2}{_1\cancel{4} \times 5}$$
$$= \frac{18}{5} = 3\frac{3}{5}$$

$$3\frac{1}{2} \div 2\frac{2}{3} = \frac{7}{2} \div \frac{8}{3}$$
$$= \frac{7}{2} \times \frac{3}{8} = \frac{7 \times 3}{2 \times 8}$$
$$= \frac{21}{16} = 1\frac{5}{16}$$

> **Key Point**
>
> Always write the 'remainder' as a fraction.

> **Quick Test**
>
> Work out:
>
> 1. $\frac{4}{5} \times \frac{5}{12}$
>
> 2. $\frac{7}{12} \div \frac{3}{7}$
>
> 3. $\frac{12}{15} \div \frac{5}{35}$
>
> 4. $5\frac{5}{6} + 3\frac{5}{12}$
>
> 5. $4\frac{3}{10} - 2\frac{1}{9}$
>
> 6. $1\frac{3}{4} + 2\frac{1}{5}$
>
> 7. $6\frac{1}{2} \div 2\frac{1}{4}$

> **Key Words**
>
> mixed number
> improper fraction

Coordinates and Graphs 1

You must be able to:

- Plot linear graphs
- Understand the components of $y = mx + c$
- Solve equations from linear graphs.

Linear Graphs

- **Coordinates** are usually given in the form (x, y) and they are used to find certain points on a graph with an x-axis and a y-axis.
- **Linear** graphs form a straight line.

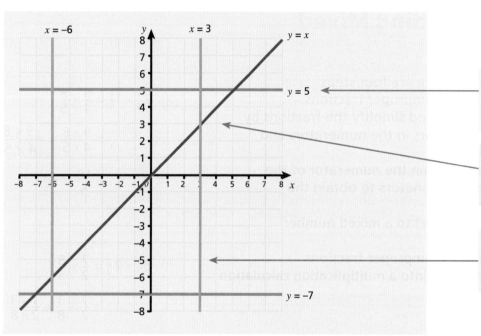

Plotting $y = 5$, we can choose any value for the x-coordinate but y must always equal 5. This line is parallel to the x-axis.

Plotting $y = x$, whatever x-coordinate you choose, the y-coordinate will be the same, e.g. (3, 3).

Plotting $x = 3$, we can choose any value for the y-coordinate but x must always equal 3. This line is parallel to the y-axis.

Graphs of $y = mx + c$

- m is the **gradient** of the graph.
- c is the **intercept** with the y-axis.
- To create the graph we substitute numbers for x and y.

Plot the graph $y = 2x + 1$

If $x = 1$ you can work out what y is as follows:
$y = 2 \times 1 + 1 = 3$

Now change x to 2, 3, ...

x	−1	0	1	2	3
y	−1	1	3	5	7

Key Point

A positive value of m will give a positive gradient. The graph will appear 'uphill'.

A negative value of m will give a negative gradient. The graph will appear 'downhill'.

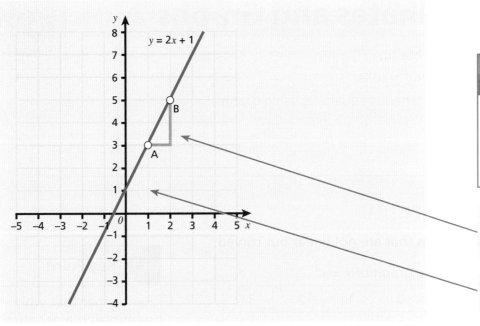

$y = 2x + 1$

When x increases by 1, y increases by 2

$2 \div 1 = 2$ so the gradient is 2

The line crosses the y-axis at 1 so the intercept is 1

- You can work out the equation of a graph by looking at the gradient and the intercept.
- The gradient can be worked out by picking two points on the graph, finding the difference between the points on both the y- and x-axes and dividing them:

$$\frac{\text{difference in } y}{\text{difference in } x} = \text{gradient}$$

- The y-intercept is where the graph crosses the y-axis.

Solving Linear Equations from Graphs

- You can use a graph to find the solution to an equation.

Using the graph $y = 2x + 1$ (above), you can find the solution to the equation $5 = 2x + 1$

First of all, plot the graph – see above. Then find where $y = 5$ on the axis.

Trace your finger across until it meets the graph and finally follow it down to read the x-axis value: $x = 2$

Quick Test

1. Complete the table below for the equation of $y = 3x - 5$, then plot the graph.

x	−2	−1	0	1	2	3
y						

2. What is the equation of a linear graph with a gradient of 5 and a y-intercept of 3?

Coordinates and Graphs 2

You must be able to:

- Plot quadratic graphs
- Solve simultaneous equations using graphs.

Drawing Quadratic Graphs

- Quadratic equations make graphs that are not linear but curved.

Key Point

Remember that when multiplying a negative by another negative, the result is a positive number.

Use a table of values to plot the graph of $y = x^2$

x	−3	−2	−1	0	1	2	3
$y = x^2$	9	4	1	0	1	4	9

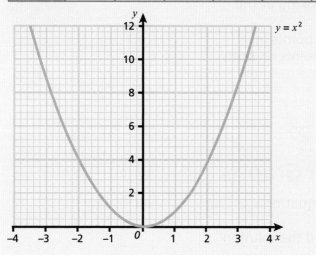

Because the equation $y = x^2$ has a power in it, this alters the graph to one that has a curve.

- Quadratics can take more complicated forms, but you still substitute a number for x to get the coordinates.

$y = x^2 + 2x + 1$

If $x = -3$, you can work out y as follows:

$y = (-3)^2 + 2(-3) + 1 = 9 + -6 + 1 = 4$

You can then work out the rest of the values of y by changing the value of x.

x	−3	−2	−1	0	1	2
y	4	1	0	1	4	9

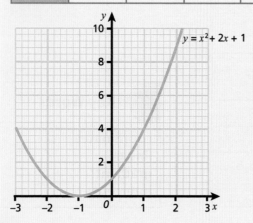

Solving Simultaneous Equations Graphically

- When you plot **simultaneous equations**, the solution to both equations can be seen at the point where the lines of the equations cross.

Solve the simultaneous equations $y = x + 2$ and $y = 8 - 2x$

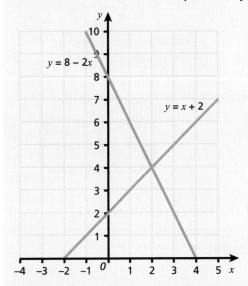

So the solution is $x = 2$ and $y = 4$

Quick Test

1. What are the gradient and y-intercepts of these equations?
   a) $y = 3x + 5$ b) $y = 6x - 7$ c) $y = -3x + 2$
2. Copy and complete the table of values for $y = x^2 + 3x + 4$

x	−3	−2	−1	0	1	2	3
y							

Key Words

quadratic equations
simultaneous equations

Review Questions

3D Shapes: Volume and Surface Area

1 Work out the surface area and the volume of this cuboid.

Do not forget the units. **[4]**

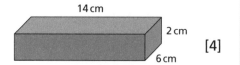

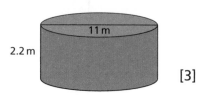

2 Work out the volume of this oil drum to the nearest whole unit and state the units.

[3]

(MR) **3** If the volume of a cube is $512\,m^3$, what is the length of the sides in centimetres? **[2]**

(MR) **4** Parveen has $1200\,cm^2$ of paper to wrap this birthday present.

Does he have enough paper? Show your working.

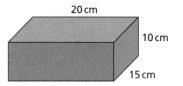

[3]

Total Marks _____ / 12

1 What is the volume and surface area (including the base) of this house? **[4]**

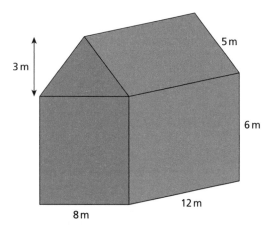

Total Marks _____ / 4

Interpreting Data

(MR) **1** Plot the following data on a graph.

Discuss any patterns, and what the graph shows.

TV viewing figures (in 1000s)	50	45	25	65	80	75	40	30	55
TV advert spend (in £1000s)	40	30	10	45	60	70	35	15	30

[4]

2 For each survey question below, state two things that could be improved.

a) Do you eat a lot of junk food?   Yes ☐   No ☐   [2]

b) How much fruit do you eat in a week?   1 ☐   2 ☐   3 ☐   4 ☐   [2]

Total Marks _____ / 8

1 Compare these two pie charts that show how two people spend their spare time each week. [3]

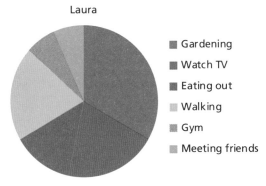

Laura

- Gardening
- Watch TV
- Eating out
- Walking
- Gym
- Meeting friends

Jules

- Gardening
- Watch TV
- Eating out
- Walking
- Gym
- Meeting friends

(MR) **2** Katya needs to spin a green to finish a game. After 12 attempts, the spinner has not landed on green.

What hypothesis might Katya suggest and how could she test it? [3]

Total Marks _____ / 6

Practice Questions

Fractions

1 Work out each calculation. Simplify your answers.

a) $\frac{4}{10} + \frac{1}{4} + \frac{2}{5} =$

b) $\frac{2}{5} + \frac{1}{8} + \frac{1}{2} =$

c) $\frac{3}{4} + \frac{2}{5} + \frac{3}{10} =$

d) $\frac{3}{4} - \frac{1}{8} =$

e) $\frac{5}{6} - \frac{1}{5} - \frac{1}{3} =$

f) $\frac{7}{9} - \frac{1}{4} =$ [6]

2 Work out each calculation. Simplify your answers.

a) $\frac{1}{8} \times \frac{2}{3} =$

b) $\frac{5}{6} \times \frac{8}{9} =$

c) $\frac{3}{10} \times \frac{1}{2} =$ [3]

3 Work out each calculation. Simplify your answers.

a) $\frac{1}{8} \div \frac{2}{3} =$ [1]

b) $\frac{1}{6} \div \frac{8}{9} =$ [2]

c) $\frac{3}{4} \div \frac{3}{7} + \frac{1}{2} =$ [2]

Total Marks _____ / 14

MR **1** Work out each calculation. Give your answer as a mixed number where appropriate.

a) $4\frac{3}{8} + 2\frac{1}{5} =$ [2]

b) $3\frac{3}{5} + 2\frac{3}{9} + 3\frac{1}{2} =$ [2]

c) $7\frac{1}{4} - 2\frac{8}{11} =$ [2]

d) $2\frac{1}{5} - 1\frac{3}{7} =$ [2]

e) $1\frac{1}{2} \times 4\frac{2}{3} =$ [2]

f) $3\frac{1}{5} \div 1\frac{2}{5} =$ [3]

Total Marks _____ / 13

Coordinates and Graphs

(PS) **1** The points in this graph are reflected in the y-axis.

Write down the coordinates of their new positions.

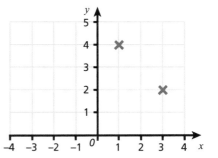

[2]

2 Draw the graph that shows the following lines:

a) $y = x$ b) $y = -x$ [2]

3 Copy and complete the table for the equation below and plot your results on a graph.

$y = -3x + 4$

[4]

x	−1	0	1	2	3
y					

Total Marks _____ / 8

1 What is the gradient and y-intercept of this graph?

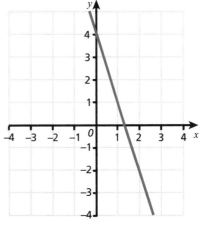

[2]

(MR) **2** Copy and complete the table of values for the quadratic equation $y = x^2 - 3x - 2$

[3]

x	−3	−2	−1	0	1	2	3
y							

Total Marks _____ / 5

Angles 1

You must be able to:

- Measure and draw angles
- Use properties of triangles to solve angle problems
- Use properties of quadrilaterals to solve angle problems
- Calculate angles at a point, on a straight line and at a right angle
- Bisect an angle.

How to Measure and Draw an Angle

- Angles are measured in **degrees**.
- A **protractor** is used to measure and draw an angle.

Remember:
- angles less than 90° are called acute angles
- angles between 90° and 180° are called obtuse angles
- angles greater than 180° are called reflex angles.

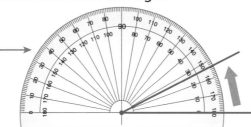

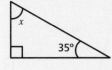

> **Key Point**
>
> There are two scales on your protractor. This is so the protractor can be used in two different directions. Always start at zero and count up.

To measure the angle of a line:
- Place the protractor with the zero line on the base line.
- The centre should be level with the point where the two lines meet.
- Counting up from zero, count the degrees of the angle you are measuring.

To draw an angle:
- Draw a base line for the angle.
- Line up your protractor, putting the centre on one end of the line.
- Count up from zero until you reach your angle, e.g. 45°
- Put a mark. Remove the protractor and draw a straight line joining the end of the base line to your point.

Angles in Triangles and Quadrilaterals

- Angles in any **triangle** add up to 180°. You can use this fact to help you work out unknown angles.
- An isosceles triangle has two equal sides and two (base) angles that are equal.
- A right-angled triangle has a 90° angle, so if you have one more angle you can work out the remaining one.

Find the size of angle x in these triangles.

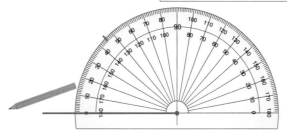

$180° - (90° + 35°) = x = 55°$

$180° - (53° \times 2) = x = 74°$

- Angles in any **quadrilateral** add up to 360°

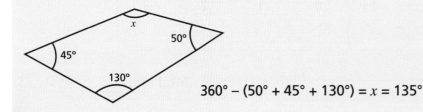

$360° - (50° + 45° + 130°) = x = 135°$

> **Key Point**
>
> Each triangle or quadrilateral will present properties that are useful when working out unknown angles.

- A parallelogram has two sets of equal angles. The opposite angles are equal.
- Only one angle is needed to be able to work out the others.

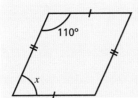

$360° - (110° \times 2) = 140°$
Now share this value equally between the remaining two angles:
$x = 140° \div 2 = 70°$

Calculating Angles

- **Angles at a point** add up to 360°
- **Angles on a straight line** add up to 180° and are called **supplementary angles**.
- **Angles in a right angle** add up to 90°
- **Vertically opposite angles** are equal.

Work out the lettered angle in each diagram.

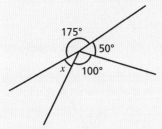

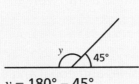

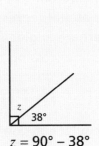

$x = 360° - 175° - 100° - 50°$
$x = 35°$

$y = 180° - 45°$
$y = 135°$

$w = 28°$

$z = 90° - 38°$
$z = 52°$

Bisecting an Angle

- **Bisect** means to cut exactly into two.

1. Open your compass.
2. Put the point of the compass on the vertex.
3. Make an arc intersecting both lines.
4. Now put the point on the first intersection and make an arc between the lines.
5. Repeat for the other intersection keeping the same radius. Make sure the arcs intersect.
6. Draw a line through the vertex and the intersection.

The vertex is where the lines meet.

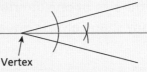

Vertex

Angles 1: Revise

Quick Test

1. Using a protractor, draw an angle of:
   a) 48°   b) 84°   c) 125°   d) 167°
2. Find the size of the marked angle in these shapes.
   a)    b) 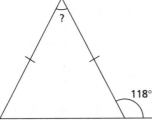   c)

Key Words

degree
protractor
triangle
quadrilateral
supplementary angles
vertically opposite angles
bisect

Angles 2

You must be able to:

- Understand and calculate angles in parallel lines
- Use properties of a polygon to solve angle problems
- Use properties of some polygons to tessellate them.

Angles in Parallel Lines

- Parallel lines are lines that run at the same angle.
- Using parallel lines and a line that crosses them, you can apply some observations to help find missing angles.
- The angles represented by ⚡ are equal. They are called **corresponding angles**.
- The angles represented by 🤍 are also equal. They are called **alternate angles**.
- The line crossing the two parallel lines is called a **transversal**.

Corresponding angles

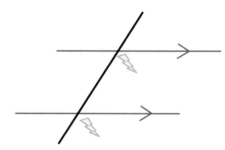

Alternate angles

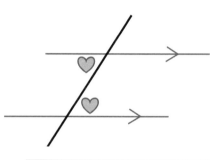

Angles in Polygons

- A **regular polygon** is a shape that has equal sides and equal angles.
- Using the fact that angles in a triangle add up to 180°, you can split any shape into triangles to help you work out the number of degrees inside that shape.

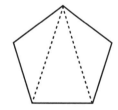

- Dividing the total number of degrees inside a regular polygon by the number of vertices will give the size of one **interior angle**.
- The sum of the **exterior angles** of any shape always equals 360°

Key Point

A polygon can be split into a number of triangles. Try this formula to speed up calculation:

(No. of sides − 2) × 180 = sum of interior angles

A regular polygon has interior angles of 140°. How many sides does it have?

 140°

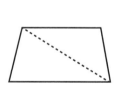 140° 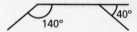 40°

The exterior angle of each part of the polygon is
180° − 140° = 40°

360° ÷ 40° = 9, so the polygon has nine sides. ◄ A polygon with nine sides is called a nonagon.

Shape	Number of sides	Sum of the interior angles
Triangle	3	180°
Quadrilateral	4	360°
Pentagon	5	540°
Hexagon	6	720°
Heptagon	7	900°
Octagon	8	1080°
Decagon	10	1440°

Properties of Triangles and Quadrilaterals

- Here is a reminder of some commonly used shapes.

Types of triangle	**Isosceles triangle** (two equal sides, two equal angles)		**Equilateral triangle** (three equal sides, three equal angles)	
Types of quadrilateral	**Square** (four equal sides, four right angles)		**Kite** (two pairs of adjacent equal sides, one pair of equal angles)	
	Rectangle (opposite sides equal, four right angles)		**Rhombus** (opposite sides parallel, four equal sides, opposite angles equal, diagonals bisect at right angles)	
	Parallelogram (opposite sides equal, opposite angles equal)		**Arrowhead** or **Delta** (two pairs of adjacent equal sides, one pair of equal angles)	
	Trapezium (one pair of parallel sides)			

Polygons and Tessellation

- Tessellation is where you repeat a shape or a number of shapes so they fit without any overlaps or gaps.

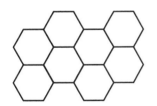

Quick Test

1. Find the size of the marked angles.

   a)  55° ?

   b) ? 112°

   c) ? 126°

2. Name a regular shape that tessellates.

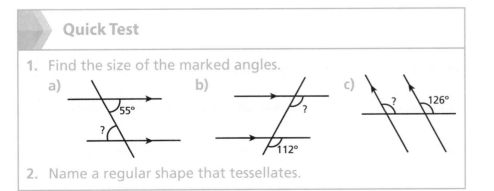

Probability 1

You must be able to:

- Recognise and use words associated with probability
- Construct and use a probability scale
- Calculate the probability of an event not occurring
- Construct and use sample spaces.

Probability Words

- Certain words are used to describe the chance of an **outcome** happening. How would you describe the chance of:
  - there being 40 days in a month?
    Impossible – there are at most 31 days in a month.
  - a student attending school tomorrow?
    Likely – it cannot be said to be certain as the student might be on school holiday or ill and not attending school.
  - rolling a 2 on a dice?
    Unlikely – there are six possible outcomes and the number 2 is only one of these.
  - taking a green sweet from a bag that only contains green sweets?
    Certain – in this case there is no other outcome possible.
  - flipping a fair coin and it landing on heads?
    An **even chance** – the outcome could be a head or a tail, two equally likely options.

> **Key Point**
>
> Try to consider the event with all the possible outcomes.
>
> Once all the possible outcomes have been considered, the word can be selected.

Probability Scale

- The **probability** of a particular outcome can be described using a **probability scale** from 0 to 1
- Probabilities are written as fractions, decimals or percentages.
- The probability of an outcome happening is written as

$$P(\text{outcome}) = \frac{\text{number of ways the outcome can happen}}{\text{total number of all possible outcomes}}$$

(unlikely)   (likely)

0	0.5	1
(impossible)	(even chance)	(certain)

Events and Outcomes

- An **event** is a set of outcomes, e.g. rolling a dice.
- An event is **fair** when the outcomes are **equally likely**.
- An event is **biased** when the outcomes are **not** equally likely.
- **Random** means each possible outcome is equally likely.

When you throw a fair coin:
$P(\text{Head})$ or $P(H) = \frac{1}{2}$ and $P(\text{Tail})$ or $P(T) = \frac{1}{2}$ ← Use this notation to save time.

A bag contains 5 red and 4 blue counters. One counter is taken from the bag at random. What is the probability it is:

a) red?   Five of the counters are red, so P(red) = $\frac{5}{9}$

b) blue?   Four of the counters are blue, so P(blue) = $\frac{4}{9}$

c) green?   There are no green counters, so P(green) = 0

- The probabilities of all possible outcomes sum to 1. In the example above, P(red or blue) = $\frac{9}{9}$ = 1

Probability of an Outcome Not Happening

- The probability of an outcome **not** happening is:
  1 – the probability of the outcome happening

A fair six-sided dice is rolled.

P(rolling a 2) = $\frac{1}{6}$

P(**not** rolling a 2) = $1-\frac{1}{6}=\frac{5}{6}$

P(rolling a 4 or a 5) = $\frac{2}{6}$

P(**not** rolling a 4 or a 5) = $1-\frac{2}{6}=\frac{4}{6}$

 There are five outcomes that are not a 2; these are 1, 3, 4, 5, 6

The probability of the weather being cloudy = 0.4

So the probability of it **not** being cloudy is 1 – 0.4 = 0.6

> **Key Point**
>
> The sum of the probabilities of all possible outcomes is 1

Sample Spaces

- A **sample space** shows all the possible outcomes of the event.
- When two or more events take place, they are **combined events**.

The sample space for throwing a coin and rolling a dice is:

H1	H2	H3	H4	H5	H6
T1	T2	T3	T4	T5	T6

This shows all the possible outcomes and helps us to calculate probabilities.

P(H1) = $\frac{1}{12}$

> **Key Words**
>
> outcome
> impossible
> likely
> unlikely
> certain
> even chance
> probability
> probability scale
> event
> fair
> equally likely
> biased
> random
> sample space
> combined events

 Quick Test

1. In the UK, how can the chance of rain in October be described?
2. Show the outcome in question 1 on a probability scale.
3. A bag contains 5 red sweets and 10 blue sweets.
   a) What is the probability of picking a red sweet?
   b) What is the probability of not picking a red sweet?
4. If it rains 0.642 of the time in the rainforest and is cloudy 0.13 of the time, what is the probability it isn't raining or cloudy?

Probability 2

You must be able to:

- Understand what mutually exclusive outcomes are
- Calculate a probability with and without a table
- Work with experimental probability
- Understand Venn diagrams and set notation.

Mutually Exclusive Outcomes

- **Mutually exclusive** outcomes are outcomes that cannot happen at the same time.
- For example, the arrow on the spinner **cannot** land on yellow and blue at the same time. Because the outcomes are mutually exclusive, P(yellow and blue) = 0
- You can calculate the probability of spinning yellow or blue as P(yellow or blue) = P(yellow) + P(blue) = $\frac{1}{4} + \frac{1}{4} = \frac{1}{2}$

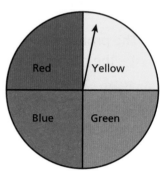

Probability Calculations

- Probabilities can be used to predict possible outcomes.
- If the outcome of one event does **not** affect the outcome of another event, they are **independent**.
- If the outcome of one event does affect the outcome of another event, they are **dependent**.
- **Conditional probability** is when the probability is affected by a previous outcome.

Key Point

When outcomes are mutually exclusive, using the word **or** implies we can add the probabilities of the outcomes together.

What is the probability of picking a green or black ball, at random, from the bag?

P(green or black) = $\frac{3}{15} + \frac{4}{15} = \frac{7}{15}$

If a red ball is removed from the bag, what is the probability the next ball is green?

There are only 14 balls left so P(green) = $\frac{3}{14}$

This is conditional probability.

This bag contains 15 balls: 3 are yellow, the rest are either red or blue. Use the table of probabilities to work out the number of each colour ball.

Yellow	Red	Blue
0.2	0.6	?

The probability that any ball is chosen is 1 so we can calculate the probability of drawing a blue ball: 1 − 0.6 − 0.2 = 0.2

There are the same number of blue as yellow as both have probability 0.2

One way of working out the actual number of each colour is to multiply the probability by the total number of balls:
15 × 0.6 = 9 (red)   0.2 × 15 = 3 (blue)

There are three times as many red as yellow as 0.6 = 3 × 0.2

Experimental Probability

- The probability of a 6 when rolling a fair six-sided dice is $\frac{1}{6}$
  If you roll the dice six times, would you definitely get a 6? You
  may not get a 6 in six rolls, however the more times you roll the
  dice the more likely you are to get closer to a probability of $\frac{1}{6}$
  This is **experimental probability**.

Hayley sat outside her school
and counted 25 cars that went
past. She noted the colour of
each car in this table.

Yellow	1
Red	6
Blue	4
Black	9
White	5

a) What is the probability of the next car going past being
white?

$$\frac{5}{25} = \frac{1}{5} = 0.2$$

b) How many black cars would you expect if 50 cars go past?

$$\frac{9}{25} = 0.36 \qquad 0.36 \times 50 = 18$$

Venn Diagrams and Set Notation

- Venn diagrams can be used to organise sets and find probabilities.

Set A is the children who like
green beans. Set B is the children who
like carrots.

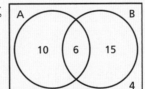

(A) = 10 + 6 = 16 children like green beans

(B) = 6 + 15 = 21 children like carrots

(A $\cup$ B) = 10 + 6 + 15 = 31 children like at least one.   ← This is the union of A and B.

(A $\cap$ B) = 6 children like both green beans and carrots.   ← This is the intersection of A and B.

(A $\cup$ B)' = 4 children like neither green beans nor carrots.   ← The symbol ' means not in A or B.

The probability that a child does **not** like green beans or
carrots = $\frac{4}{35}$

Fractions

1 Work out each calculation. Simplify your answers where possible.

a) $\frac{2}{5} + \frac{1}{10} =$

b) $\frac{7}{12} + \frac{1}{4} =$

c) $\frac{1}{6} + \frac{1}{5} =$

d) $\frac{2}{7} + \frac{3}{10} =$

e) $\frac{8}{9} - \frac{1}{3} =$

f) $\frac{7}{11} - \frac{1}{2} =$

g) $\frac{9}{10} - \frac{2}{3} =$

[9]

2 Work out each calculation. Simplify your answers where possible.

a) $\frac{4}{9} \times \frac{1}{5} =$

b) $\frac{3}{7} \times \frac{3}{10} =$

c) $\frac{5}{12} \times \frac{2}{3} =$

d) $\frac{2}{9} \div \frac{1}{4} =$

e) $\frac{4}{5} \div \frac{6}{11} =$

[7]

(MR) **3** Sally won some money in the lottery. She gave $\frac{2}{5}$ to her husband and $\frac{1}{4}$ to her daughter.

What fraction did she keep?

[3]

(MR) **4** Kamil was baking a cake. His recipe was $\frac{4}{9}$ flour, $\frac{1}{3}$ sugar and butter, and the rest an equal split of chocolate and eggs.

What fraction does the chocolate part represent?

[3]

Total Marks _____ / 22

1 Change the following mixed numbers to improper fractions.

a) $8\frac{5}{9}$

b) $3\frac{2}{7}$

c) $1\frac{3}{11}$

[3]

2 Use a calculator to work out the following.

a) $1\frac{3}{4} \times 2\frac{1}{2}$

b) $4\frac{1}{2} \div 2\frac{2}{3}$

[2]

Total Marks _____ / 5

Coordinates and Graphs

1 Copy and complete the arrows and the table below, then plot the equation $y = 2x - 4$ on a graph.

x	−1	0	1	2	3
y					

[4]

2 Copy and complete the table of values for the equation $y = x^2 + 5x + 1$

x	−3	−2	−1	0	1	2	3
y							

[3]

Total Marks _____ / 7

(MR) 1 Will the graph $y = x^2 - 4x + 6$ have the coordinates (3, 3)? Justify your answer.　[2]

(MR) 2 Find the solutions for the following simultaneous equations graphically.

$y = 4x + 2$

$y = -2x + 5$ 　[3]

(MR) 3 Match the equations with the same gradients:

A $y = 2x + 12$

B $y = 12 - 3x$

C $y = 4x - 9$

D $y = -3x + 5$

E $2y = 4x + 2$

F $y = 19 + 4x$ 　[3]

Total Marks _____ / 8

Angles

1. Find the size of the marked angle in this isosceles triangle.

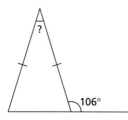

106°

[2]

(PS) 2. Find the size of the angles marked x and y in each diagram.

a)

134°

x y

b)

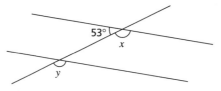

53°

x

y

[4]

3. What is the name of a polygon with 10 sides? [1]

(PS) 4. If each interior angle of a regular polygon is 150°, how many sides does it have? [2]

Total Marks _____ / 9

1. Calculate the size of the lettered angles in each of these shapes.

a)

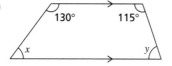

130° 115°

x y

[2]

b)

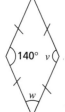

140° v

w

[2]

Total Marks _____ / 4

Probability

① Ranjeet drops a paper cup a number of times. He finds the probability of it landing upside-down is 0.65

What is the probability of it **not** landing upside-down? [2]

(MR) ② A dice has been rolled 50 times and the score recorded in the frequency table below.

a) Copy and complete the table. Give the probabilities as fractions. [3]

Number	Frequency	Estimated probability
1	5	
2	8	
3	7	
4	7	
5	8	
6	15	
Total		

b) Use the results in your table to work out the estimated probability (as fractions) of getting:

i) the number 6   ii) an odd number   iii) a number greater than 4 [3]

c) Do you think the dice is fair? Give a reason for your answer. [2]

Total Marks .... / 10

① Bradley has a pack of coloured cards. The table shows the probability of each colour being chosen.

Yellow	Red	Blue	Green	Pink
0.26	0.18	0.09		0.15

a) Copy and complete the table to show the probability of choosing a green card. [1]

b) What is the probability that a card is chosen that is not yellow? [2]

c) What is the probability that Bradley chooses a blue or green card? [1]

Total Marks .... / 4

Quick Recall Quiz

Fractions, Decimals and Percentages 1

You must be able to:

- Convert between a fraction, decimal and percentage
- Order fractions, decimals and percentages
- Calculate a fraction of a quantity and a percentage of a quantity
- Compare quantities using percentages.

Fractions, Decimals and Percentages

- The table below shows how fractions, decimals and percentages are used to represent the same amount:

Picture	Fraction	Decimal	Percentage
$\frac{1}{4}$	$\frac{1}{4}$	0.25	25%
$\frac{1}{2}$	$\frac{1}{2}$	0.5	50%
$\frac{3}{4}$	$\frac{3}{4}$	0.75	75%

> **Key Point**
>
> Percent means 'out of 100'

Carrying out Conversions

- For conversions you do not know automatically, use the rules below.

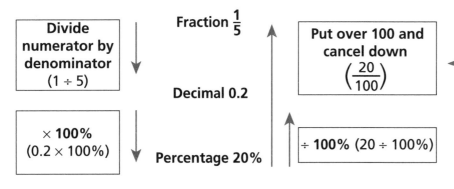

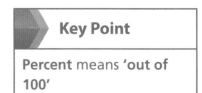

Ordering Fractions, Decimals and Percentages

- To order fractions, decimals and percentages, change them all to the same format.

Put 0.25, $\frac{2}{5}$ and 20% in order from smallest to largest.

Changing them all to decimals gives $\frac{2}{5} = 0.4$ and 20% = 0.2

0.2 < 0.25 and 0.25 < 0.4, so correct order is 20%, 0.25, $\frac{2}{5}$

Fractions of a Quantity

- To find a fraction of a **quantity** without a calculator, divide by the **denominator** and multiply by the **numerator**.

  Find $\frac{2}{3}$ of £120

  $120 \div 3 \times 2 = £80$

- To find a fraction of a quantity with a calculator, use the fraction button . Calculators can differ, so find out how yours works with fractions.

  Find $\frac{2}{5}$ of £200

  2 $5 \times 200 = £80$

Percentages of a Quantity

- To find a percentage of a quantity without a calculator:

  Find 20% of £60

  10% of £60 $= 60 \div 10 = £6$
  $20\% = £6 \times 2 = £12$

  20% is two lots of 10%

- To find a percentage of a quantity with a calculator:

  Find 20% of £60

  $= 20\% \times £60$
  $= 20 \div 100 \times 60 = £12$

  Or $20\% = 0.2$
  so 20% of £60 is
  $0.2 \times £60 = £12$

  Change the percentage to a decimal.

  Remember: as 120% > 100%, your answer will be bigger than £60

  Find 120% of £60

  $= 120\% \times £60$
  $= 120 \div 100 \times 60 = £72$

  Or $120\% = 1.2$
  so 120% of £60 is
  $1.2 \times £60 = £72$

- To compare two quantities using percentages:

  Which is bigger, 30% of £600 or 25% of £700?

  10% of £600 $= 600 \div 10 = £60$ $\quad$ 50% of £700 $= 700 \div 2 = £350$
  30% of £600 $= 60 \times 3 = £180$ $\quad$ 25% of £700 $= 350 \div 2 = £175$

  So 30% of £600 is bigger, by £5

> **Key Point**
>
> Useful percentages to know:
>
> $50\% \rightarrow \div 2$
>
> $10\% \rightarrow \div 10$
>
> $1\% \rightarrow \div 100$
>
> 'of' means '×'

Quick Test

1. Change $\frac{7}{20}$ to a decimal and a percentage.
2. Change 36% to a fraction in its simplest form.
3. Work out $\frac{2}{5}$ of £70
4. Find 35% of $140

Key Words

fraction
decimal
percentage
quantity

Fractions, Decimals and Percentages 2

You must be able to:

- Increase or decrease a quantity by a percentage
- Work out a percentage change
- Find one quantity as a percentage of another
- Work out simple interest and tax
- Work out the original amount from a given percentage.

Increasing and Decreasing Quantities by a Percentage

- To **increase** or **decrease** a quantity by a percentage, add on or subtract the percentage you have found.
- You can also use a **multiplier**.

A calculator is priced at £12 but there is a discount of 25%
Work out the reduced price of the calculator.

25% of £12 is one-quarter of £12 = £3
Reduced price = £12 – £3 = £9

> £3 is the discount so **'take it away'** to get the final answer.

Using a multiplier:
A reduction of 25% means you are left with 75%, and 75% = 0.75, so the multiplier is 0.75
0.75 × £12 = £9

A laptop computer costs £350 plus tax at 20%
Work out the actual cost of the laptop.

20% of £350 is 20% × £350 = 20 ÷ 100 × 350
= £70

Actual cost = £350 + £70 = £420

> £70 is the tax so **'add it on'** to get the final answer.

Using a multiplier:
An increase of 20% means you pay 120%, and 120% = 1.2
1.2 × £350 = £420

> So the multiplier is 1.2

Percentage Change

- You can work out the percentage change using

$$\text{Percentage change} = \frac{\text{change}}{\text{original}} \times 100\%$$

In a sale, a coat is reduced from £50 to £30
Work out the percentage reduction.

Percentage reduction is $\frac{20}{50} \times 100\% = 40\%$

The number of students in a class increases from 25 to 30. Work out the percentage increase.

Percentage increase is $\frac{5}{25} \times 100\% = 20\%$

One Quantity as a Percentage of Another

Jane gets 18 out of 20 in a test. What percentage is this?

With a calculator:
$$\frac{18}{20} \times 100\%$$
$$= 18 \div 20 \times 100\%$$
$$= 90\%$$

Without a calculator:
$$\frac{18}{20} = \frac{90}{100} = 90\%$$
(× 5)

> Make the fraction 'out of' 100

Simple Interest and Tax

- Find the **interest** for one year then multiply by the number of years.

> £200 is put in a savings account earning 5% simple interest
> per year. How much is in the account after two years?
>
> $$10\% \text{ of } £200 = 200 \div 10 = £20$$
> $$5\% = 20 \div 2 = £10$$
>
> Interest after two years is £10 × 2 = £20, so £220 in the account.

← This is the interest for one year.

> A bank charges 8% simple interest per year on loans. Work
> out the amount of interest on a three-year loan of £2000
>
> $$8\% \text{ of } £2000 = 8\% \times £2000$$
> $$= 8 \div 100 \times 2000 = £160$$
>
> Interest after three years will be £160 × 3 = £480

← This is the interest for one year.

- You have to pay tax on money you earn, called **income tax**, and
  also on some things you buy, called **value added tax (VAT)**.

> A man earns £30 000 per year. The first £12 500 is tax free. He
> pays 20% income tax on the rest. How much does he pay?
>
> He pays tax on £30 000 – £12 500 = £17 500
> 20% of 17 500 = 20 ÷ 100 × 17 500 = £3500

> A plumber charges £60 per hour plus VAT. VAT is 20%
> How much does he charge including VAT?
>
> VAT is 20% so 100% + 20% = 120% = 1.2
> 1.2 × £60 = £72

← So the multiplier is 1.2

Reverse Percentages

- In these questions, you are given the cost **after** the increase/
  decrease and will have to find the original cost.

> A car's value decreases by 20% to £5000. Find its original price.
>
> $$80\% = £5000$$
> $$1\% = £5000 \div 80 = £62.50 \quad \text{So } 100\% = £6250$$
> The original price = £6250

The **decrease** of 20% means that
£5000 represents 80% of the
original value.

If it was an **increase** of 20% then
you would write 120% = £5000

Quick Test

1. A TV costing £450 is reduced by 10%. What is its sale price?
2. An £80 000 house rises in value by 15%. What is its new value?
3. Anna gets $\frac{21}{25}$ in a test. What percentage is this?
4. Write 24 cm as a percentage of 30 cm.
5. I save £400 at simple interest of 5%. How much will I have
   after three years?
6. Find the original price of a house that has increased in value
   by 10% to £165 000

Key Words

increase
decrease
multiplier
interest
income tax
value added tax (VAT)

Quick Recall Quiz

Algebra

Equations 1

You must be able to:

- Find an unknown number
- Solve a simple equation
- Solve an equation with unknowns on both sides
- Apply the inverse of an operation.

Finding Unknown Numbers

- The **unknown** number is usually given as a letter or symbol.
- Applying the **inverse** operations helps you to find the unknown number.

$n + 4 = 13$ or $\square + 4 = 13$ ←—— Both ways mean **'something + 4 = 13'**

Take 4 away from 13

n or $\square = 13 - 4$ ←—— -4 is the 'inverse' or 'opposite' of $+4$

n or $\square = 9$ (check: $9 + 4 = 13$)

$x - 4 = 13$ or $\bigcirc - 4 = 13$ ←—— Both ways mean **'something − 4 = 13'**

Add 4 to 13

x or $\bigcirc = 13 + 4$ ←—— $+4$ is the 'inverse' or 'opposite' of -4

x or $\bigcirc = 17$ (check: $17 - 4 = 13$)

- The same idea applies to multiplying and dividing.

$3n = 12$ ←—— $3 \times$ something $= 12$

$n = 12 \div 3$ ←—— $\div 3$ is the inverse of $\times 3$

$n = 4$ (check: $3 \times 4 = 12$)

$\dfrac{n}{3} = 4$ ←—— Something $\div 3 = 4$

$n = 4 \times 3$ ←—— $\times 3$ is the inverse of $\div 3$

$n = 12$ (check: $12 \div 3 = 4$)

Solve $\dfrac{3}{4} x = 24$

$3x = 24 \times 4$

$x = \dfrac{24 \times 4}{3} = \dfrac{96}{3} = \dfrac{\overset{8}{\cancel{24}} \times 4}{\cancel{3}_1}$

$x = 32$ (check: $32 \div 4 \times 3 = 24$)

- Now you can apply inverse operations to **solve** more difficult equations.

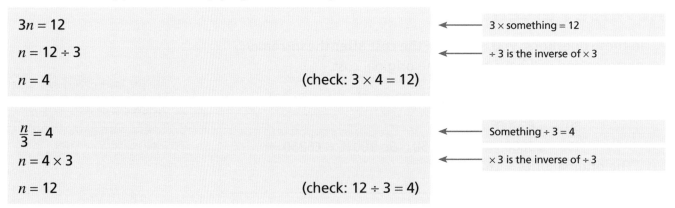

Key Point

Operation	Inverse
$+$	$-$
$-$	$+$
$\times$	$\div$
$\div$	$\times$

82   **KS3 Maths Revision Guide**

Solving Equations

- Remember to think of the letter as 'something'.

Solve the equation $2y + 3 = 15$

This simply means 'something' $+ 3 = 15$ 'Something' must be 12

So $2y = 12$ means $2 \times$ 'something' $= 12$ ⟵ 'Something' must be 6

So $y = 6$

Now look with the **inverses**:

$$2y + 3 = 15$$
$$(- 3) \quad 2y = 12$$
$$(\div 2) \quad y = 6$$

Key Point

Remember to do the same to both sides of the equation.

Equations with Unknowns on Both Sides

- An equation may have an unknown number on both sides of the equals sign.

Solve the equation $5x - 2 = 3x + 5$

$$5x - 2 = 3x + 5$$ ⟵ Do the same to both sides.
$$(- 3x) \quad 2x - 2 = 5$$ ⟵ Subtract $3x$ from both sides so that the x term is on the left-hand side only.
$$(+ 2) \quad 2x = 7$$
$$(\div 2) \quad x = 3.5$$ ⟵ Now do the inverses.

Solve the equation $3x + 5 = 5x - 4$

$$3x + 5 = 5x - 4$$
$$(- 3x) \quad 5 = 2x - 4$$
$$(+ 4) \quad 9 = 2x$$
$$(\div 2) \quad 4.5 = x$$ ⟵ This is the same as $x = 4.5$

Quick Test

1. What is the inverse of $\times 6$?
2. If $- 5 = 7$, what is the value of ⬤?
3. If $6n = 30$, what is the value of n?
4. Solve the equation $3y - 2 = 13$
5. Solve the equation $3x + 7 = 2x - 2$

Key Words

unknown
solve

Equations 2

You must be able to:

- Solve equations with fractions or negative numbers
- Set up and solve an equation
- Apply the inverse of an operation.

Solving More Complex Equations

- An equation may include a **negative** of the unknown number.
- The unknown number may be part of a fraction or inside **brackets**.

Solve the equation $3x + 1 = 11 - 2x$

$$3x + 1 = 11 - 2x$$

$(+ 2x) \quad 5x + 1 = 11$

$(- 1) \quad\quad 5x = 10$

$(÷ 5) \quad\quad x = 2$

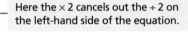
Adding $2x$ to both sides so the x term is on the left-hand side only.

Solve the equation $\dfrac{3x + 1}{2} = 8$

$$\dfrac{3x + 1}{2} = 8$$

$(× 2) \quad 3x + 1 = 16$

$(- 1) \quad\quad 3x = 15$

$(÷ 3) \quad\quad x = 5$

Here the × 2 cancels out the ÷ 2 on the left-hand side of the equation.

Solve the equation $3(2x - 1) = 4(x + 2)$

Multiply out the brackets first then solve in the usual way:

$$3(2x - 1) = 4(x + 2)$$

$$6x - 3 = 4x + 8$$

$(- 4x) \quad 2x - 3 = 8$

$(+ 3) \quad\quad 2x = 11$

$(÷ 2) \quad\quad x = 5.5$

Key Point

Remember to multiply everything inside the bracket by the number outside the bracket.

Solve the equation $\dfrac{5x}{x + 2} = 4$

(Multiply both sides by $x + 2$) $\quad 5x = 4(x + 2)$

(Multiply out) $\quad 5x = 4x + 8$

$(- 4x) \quad\quad\quad x = 8$

Setting Up and Solving Equations

- You will have to follow a set of instructions in a given order.
- Usually you only have to ×, ÷, +, − or square.
- If you have to multiply 'everything' then remember to use brackets.

A number is **doubled**, then 5 is added to the total and the result is 11. What was the original number?

The words	The algebra
A number	n
doubled	$2n$
add 5	$2n + 5$
the result is 11	$2n + 5 = 11$

Solve in the usual way to find the original number was 3

Three boys were paid £10 per hour plus a tip of £6 to wash some cars. They shared the money and each got £12. How many hours did they wash cars for?

£10 × number of hours

£6 tip

Set up an equation: $\dfrac{10x + 6}{3} = 12$ each got

shared by 3 boys

Now solve the equation:

$(\times 3)$ $10x + 6 = 36$
$(- 6)$ $10x = 30$

The boys worked for 3 hours. $(\div 10)$ $x = 3$

x is a number. The ratio $2x : 3 = x + 21 : 6$

Set up an equation and solve it to work out the value of x.

$$\frac{2x}{3} = \frac{x + 21}{6}$$

$(\times 6)$ $\dfrac{2x}{3} \times 6 = \dfrac{x + 21}{6} \times 6$

(Simplify) $4x = x + 21$

$(- x)$ $3x = 21$

$(\div 3)$ $x = 7$

Quick Test

1. Solve the equation $3x - 4 = 11$
2. Solve the equation $2x + 3 = 12 - x$
3. Solve the equation $2(x + 2) = 2(3x - 2)$
4. A number is multiplied by 3 and then 8 is subtracted. The result is 25. What is the solving number?
5. A chicken is roasted for 60 minutes for every kilogram, plus an extra 20 minutes. If the chicken took 140 minutes to cook, how heavy was it?

Key Words

negative
brackets
double

Review Questions

Angles

1 Find the size of the lettered angles in the parallel line diagrams below.

a)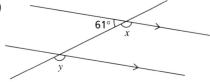

b)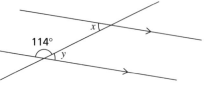

[4]

2 What do the total interior angles add up to in a nonagon? [1]

3 If each interior angle of a regular polygon is 160°, how many sides does it have?

[3]

Total Marks _____ / 8

---

1 Explain why regular pentagons cannot be used on their own for tessellation.

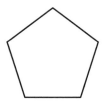

[3]

(MR) **2** Tessellate the following shape at least six times.

[2]

Total Marks _____ / 5

Probability

1 Leanne runs an ice-cream van. At random, she chooses which kind of sprinkles to put on the ice-creams. The table below shows the sprinkles Leanne chose on Sunday. 🖩

Sprinkles	Frequency	Probability
Chocolate	19	
Hundreds and thousands	14	
Strawberry	7	
Nuts	10	

a) Copy the table and complete the experimental probabilities. [2]

b) What was the probability of getting either nuts or chocolate sprinkles? [2]

2 If the probability of winning a raffle prize is 0.47, what is the probability of not winning a raffle prize? 🖩 [1]

3 a) Copy and complete the table below. 🖩

Sales destination	Probability of going to destination
London	0.26
Cardiff	0.15
Chester	0.2
Manchester	

[1]

b) Which is the least likely destination to travel to for sales? [1]

Total Marks _____ / 7

1 Yvonne works in insurance. The probability that Yvonne gets a claim from a call is 0.68

On Monday she gets 325 calls. What is the estimated number of claims? [2]

(MR) **2** Patrick is a baker. On Monday he made 250 bread rolls. However, Patrick's oven is slightly faulty and burns 0.14 of them. How many rolls were good on Monday? [2]

Total Marks _____ / 4

Fractions, Decimals and Percentages

(PS) 1 Copy and complete the following table of equivalent values.

Fraction	Decimal	Percentage
$\frac{3}{5}$		
		55
	0.32	
$\frac{3}{100}$		

[4]

(PS) 2 Work out: 🔲

a) $\frac{2}{3}$ of £15 [2]

b) $\frac{3}{7}$ of £210 [2]

c) $\frac{4}{9}$ of £27 [2]

Total Marks _____ / 10

(PS) 1 Work out: 🔲

a) 15% of 80 cm [2]

b) 35% of 160 m [3]

c) 5% of £70 [2]

(PS) 2 A jacket costing £75 is reduced by 20% in a sale. What is the sale price of the jacket? 🔲 [3]

(FS) 3 Kim puts £300 into a savings account. She will receive 5% simple interest each year.

How much will she have in the bank after the following?

a) Two years [3]

b) Five years [3]

Total Marks _____ / 16

Equations

PS **1** Solve these equations:

a) $2x - 5 = 3$ [2]

b) $3x + 1 = x + 7$ [2]

c) $2(2x - 3) = x - 3$ [2]

d) $\dfrac{3x + 5}{4} = 5$ [2]

PS **2** A number is multiplied by 3, then 2 is added to the total. The result is 11

What was the original number? [2]

3 At Anne's party there were 48 cans of drink. Everybody at the party had 4 cans each.

How many people were at the party? [2]

Total Marks _____ / 12

---

1 Solve the equation $3(x + 1) = 2 + 4(2 - x)$ [3]

2 Solve the equation $5(2a + 1) + 3(3a - 4) = 4(3a - 6)$ [3]

3 x is a number. $x : 5 = x + 6 : 15$

Work out the value of x. [4]

4 y is a number. $14 : y = 7 : y - 4$

Work out the value of y. [4]

MR **5** The rectangle and the triangle have the same perimeter.

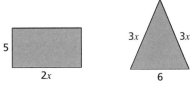

Write down an equation and solve it to find the value of x. [4]

Total Marks _____ / 18

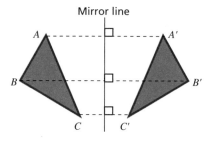

Symmetry and Enlargement 1

You must be able to:

- Reflect a shape
- Translate a shape
- Find the order of rotational symmetry
- Rotate a shape
- Enlarge a shape.

Reflection and Reflectional Symmetry

- Reflect each point one at a time.
- Use a line that is **perpendicular** to the mirror line.
- Make sure the **reflection** is the same distance from the mirror line as the original shape.
- A shape has reflectional symmetry if you can draw a mirror line through it.

Mirror line

Translation

- Translation moves a shape left/right (x) and/or up/down (y).
- The translation is described using a column vector $\begin{pmatrix} x \\ y \end{pmatrix}$

> **Key Point**
>
> Perpendicular means at right angles (90°) to.
>
> Check the position of your reflection by placing a mirror along the mirror line.

Rotational Symmetry

- A shape has rotational symmetry if it looks exactly like the original shape when it is **rotated**.
- The **order of rotational symmetry** is the number of ways the shape looks the same.
- To rotate a shape you need to know:
  - The **centre of rotation**
  - The direction of rotation
  - The number of degrees to rotate it.

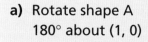

Order 1   Order 2   Order 3   Order 4

a) Rotate shape A 180° about (1, 0)

b) Translate shape A by vector $\begin{pmatrix} -4 \\ 1 \end{pmatrix}$

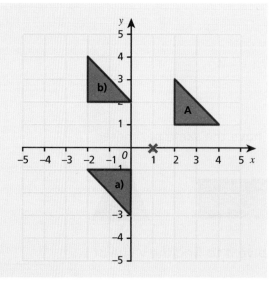

Enlargement

- To draw an **enlargement** you need to know two things:
  - How much bigger/smaller to make the shape. This is called the **scale factor**.
  - Where you will enlarge the shape from. This is called the **centre of enlargement**.
- Remember that the original and enlarged shapes are **similar** (the same shape but a different size).

Enlarge shape A by a scale factor of 2 from the point (3, 4)

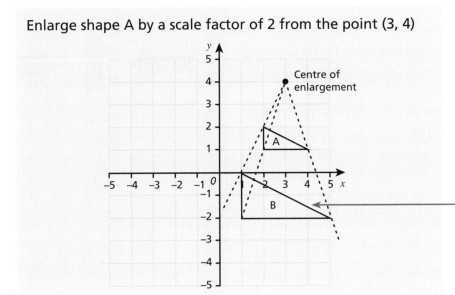

Centre of enlargement

Enlarge every side of the shape.

- Use **rays** to check the position of your enlargement. They will touch corresponding corners of the shape.

Quick Test

1. What is the order of rotational symmetry of this shape?

2. a) Reflect this shape across the mirror line.
   b) Rotate the original shape 180° about the corner A.
   c) Rotate the shape 90° clockwise about the corner B.
3. Enlarge this shape by a scale factor of 3

1cm

2cm

Symmetry and Enlargement 2

You must be able to:

- Recognise congruent shapes
- Interpret a scale drawing
- Work out side lengths in similar shapes
- Convert between units of measure
- Carry out ruler and compass constructions.

Congruence

- Congruence simply means shapes that are exactly the same.
- These arrow shapes are **congruent** – they have the same size and the same shape.
- Triangles are congruent to each other if:
  - three pairs of sides are equal (SSS)
  - two pairs of sides and the angle between them are equal (SAS)
  - two pairs of angles and the side between them are equal (ASA)
  - both triangles have a right angle, the hypotenuses are equal and one pair of corresponding sides is equal (RHS).

Same size   Same shape

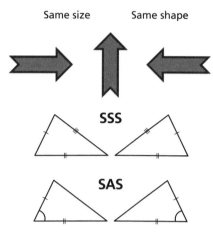

SSS

SAS

ASA

RHS

Scale Drawings

- A scale drawing is one that shows a real object with accurate dimensions, except they have all been reduced or enlarged by a certain amount (called the **scale**).
- Similar shapes are enlarged by the same scale factor, but the angles stay the same. All sides must be multiplied by the same value.
- A scale of 1 : 10 means in the real world the object would be 10 times bigger than in the drawing.
- We use scale drawings to represent real objects.

Shape and Ratio

- You can use **ratios** to work out the 'real' size of an object. The scale is given as a ratio with the smaller **unit** first.

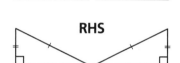

Jessie's Room

desk

wardrobe

bed

length

Scale: 1 inch = 3 feet

Here you would have to measure each side in inches then **multiply by 3** to get the real length in feet.

For every 1 cm you measure in the picture, multiply by 160 to get the real size. Then convert to metres.

Estimate the height of this house using the scale of 1 : 160

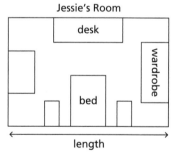

5 cm

6 cm

Key Point

Scales are given as a ratio, usually 1 : n where n is what you multiply by.

The height of the house is 5 cm so in real life it is:

$5 \times 160 = 800 \text{ cm} = 8 \text{ m}$

The width of the house is 6 cm so in real life it is:

$6 \times 160 = 960 \text{ cm} = 9 \text{ m } 60 \text{ cm}$

Remember: 100 cm = 1 m

Constructions

- You need to learn these constructions.

The **perpendicular bisector** of line AB.	A **perpendicular from a given point** to the line AB.
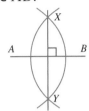	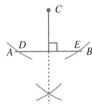
1. Open your compass to more than half AB. 2. Put the compass point on A. 3. Draw an arc extending past the centre both above and below the line. 4. Put the compass point on B and repeat Step 3. 5. Draw a line through the two points where the arcs intersect. XY is the perpendicular bisector.	1. Put the compass point on C. 2. Draw an arc on both sides of the line. 3. Keeping the radius the same, put the compass point on E. 4. Draw an arc. 5. Put the compass point on D. 6. Draw an arc. 7. Join C to the point where the arcs cross.

- A **locus** is a path of a point that moves, according to a rule, so the perpendicular bisector gives the locus of all points that are the same distance from A as B.

Quick Test

1. Draw three congruent shapes.
2. Which shape is congruent to shape A?

A   B   C   D

3. Estimate the length, in metres, of the boat. Scale: 1 : 80

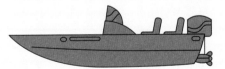

4. Which of these shapes are similar?

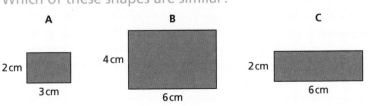

A   2 cm   3 cm   B   4 cm   6 cm   C   2 cm   6 cm

5. Draw two shapes that are similar.

> **Key Words**
>
> congruent
> scale
> ratio
> units
> perpendicular bisector
> locus

Ratio and Proportion 1

You must be able to:

- Change confidently between related standard units
- Understand what ratio means
- Simplify a ratio
- Link ratios to fractions and fractions to ratios.

Changing Units

- You need to know some conversions for standard units.

Time	Length	Area	Volume	Capacity	Mass
1 minute = 60 seconds	1 cm = 10 mm	1 cm² = 100 mm²	1 cm³ = 1000 mm³	1 litre = 1000 ml	1 kg = 1000 g
1 hour = 60 minutes	1 m = 100 cm	1 m² = 10 000 cm²	1 m³ = 1 000 000 cm³	1 litre = 100 cl	1 tonne = 1000 kg
	1 km = 1000 m	1 km² = 1 000 000 m²			

Introduction to Ratios

- Ratio is a way of showing the relationship between two numbers.

> Here is a ratio table with a missing number. Work out the missing number.
>
> | 3 | 12 |
> |---|----|
> | 18 | |
>
> The horizontal multiplier is × 4   3 × 4 = 12
> The vertical multiplier is × 6   3 × 6 = 18
>
> So the missing number is 18 × 4 or 12 × 6 = 72

- Ratios can be used to compare costs, masses and sizes.

> What is the ratio of black tiles to blue tiles?
>
> The ratio of black tiles to blue tiles is 5 : 9

> On the deck of a boat there are 2 women and 1 man. There are also 5 cars and 2 bicycles.
>
> The ratio of men to women is 1 to 2, written 1 : 2
> The ratio of women to men is 2 to 1, written 2 : 1
> The ratio of cars to bicycles is 5 to 2, written 5 : 2
> The ratio of bicycles to cars is 2 to 5, written 2 : 5

> **Key Point**
>
> 'to' is replaced with ':'

Ratios and Fractions

- Ratios can also be written as fractions.

In the previous example:

$\frac{2}{3}$ are women and $\frac{1}{3}$ are men.

$\frac{5}{7}$ of the vehicles are cars and $\frac{2}{7}$ of the vehicles are bicycles.

Simplifying Ratios

- The following ratios are **equivalent**. The relationship between each pair of numbers is the same:

10 : 20

3 : 6

2 : 4

1 : 2   This is a **simpler** way of writing the ratio 10 : 20

- You can simplify a ratio if you can divide by a common factor.
- When a ratio cannot be simplified, it is said to be in its **lowest terms**.

Simplify the ratio 30 : 100

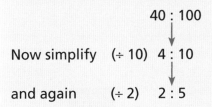 30 ÷ 10   3 : 10   100 ÷ 10   ←   Divide both numbers by 10

Write 40p to £1 as a ratio in its lowest terms.

First get the units the same: in pence, £1 is 100p

40 : 100

Now simplify   (÷ 10) 4 : 10

and again   (÷ 2)   2 : 5   ←   This is now in its lowest terms.

The angles of a triangle are 20°, 60° and 100°

What is the ratio of the angles in its lowest terms?

20° : 60° : 100°

(÷ 20) 1 : 3 : 5   ←   This is now in its lowest terms.

Quick Test

1. Look at this pattern of grey and green tiles:

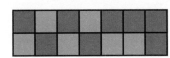

   a) Write down the ratio of green tiles to grey tiles.
   b) Write down the ratio of grey tiles to green tiles.
2. Write the following ratios in their lowest terms.
   a) 3 : 9   b) 28 : 4   c) 25 cm : 1 m

Key Word

lowest terms

Ratio and Proportion 2

Quick Recall Quiz

You must be able to:

- Share in a given ratio
- Solve problems involving direct and inverse proportion
- Use the unitary method.

Sharing Ratios

- Sharing ratios are used when a total amount is to be **shared** or **divided** into a given ratio.

> Share £200 in the ratio 5 : 3
>
> Add the ratio to find how many parts there are.
>
> $\quad$ 5 + 3 = 8 parts
>
> Divide £200 by 8 to find out how much 1 part is.
>
> $\quad$ 200 ÷ 8 = 25
>
> $\quad$ 1 part is £25
>
> Now multiply by each part of the ratio.
>
> $\quad$ 5 × £25 = £125
>
> $\quad$ 3 × £25 = £75
>
> £200 shared in the ratio 5 : 3 is £125 : £75

Key Point

Divide to find one, then multiply to find all.

> A sum of money is shared in the ratio 2 : 3
>
> If the smaller share is £30, how much is the sum of money?
>
> 2 parts = £30 so 1 part = £30 ÷ 2 = £15
>
> 3 parts = £15 × 3 = £45
>
> The sum of money = £30 + £45 = £75

Direct and Inverse Proportion

- Two quantities are in **direct proportion** if their ratios stay the same as the quantities get larger or smaller.

> If the ratio of teachers to students in one class is 1 : 30, then three classes will need 3 : 90

- Graphs of direct proportion are always this shape.
- A straight line is drawn through the points and the graph passes through the origin.
- Using algebra, $s = 30t$, where s is the number of students and t is the number of teachers.

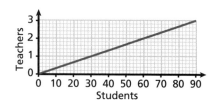

- Two quantities are in **inverse proportion** if as one quantity increases, the other decreases at the same rate.
- Speed and time are in inverse proportion. As speed increases, time decreases.

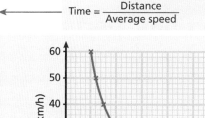

$$\text{Time} = \frac{\text{Distance}}{\text{Average speed}}$$

Suppose you go on a car journey of 60 km. The time it takes depends on the average speed of the car. The table shows some times for the 60 km journey.

Average speed, s (km/h)	10	20	30	40	50	60
Time, t (h)	6	3	2	1.5	1.2	1

Using algebra, $st = 60$
This produces a **reciprocal graph**.

Graphs of inverse proportion are always this shape. A smooth curve is drawn through the points.

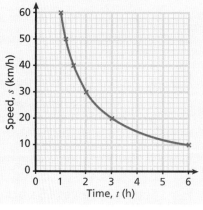

Six men build a wall in 3 days. How long will four men take?

The assumption is that they all work at the same rate.

1 man takes $3 \times 6 = 18$ days
4 men take $18 \div 4 = 4.5$ days

One man will take longer to build the wall.

Four men will take $\frac{1}{4}$ of the time taken by one man.

Using the Unitary Method

- Using the unitary method, find the value of **one unit** of the quantity before working out the required amount.

Five loaves cost £4.25. How much will three loaves cost?

Remember: divide to find one, then multiply to find all.

One loaf costs £4.25 $\div$ 5 = 85p

Three loaves will cost 85p $\times$ 3 = £2.55

This recipe for making apple pie serves four people:

200g flour   200g butter   50g sugar   8 large apples

Change these quantities to a recipe for 10 people.

Divide to find one, then multiply to find all.

Flour = 200g $\div$ 4 $\times$ 10 = 500g   Sugar = 50g $\div$ 4 $\times$ 10 = 125g

Butter = 200g $\div$ 4 $\times$ 10 = 500g   Large apples = 8 $\div$ 4 $\times$ 10 = 20

All the amounts have increased in proportion (by $2\frac{1}{2}$ times in this example).

Quick Test

1. Share 40 sweets in the ratio 2 : 5 : 1
2. £360 is divided between Sara and John in the ratio 5 : 4 How much did each person receive?
3. Work out the missing number in the ratio 4 : 5 = ? : 35
4. If six books cost £27, how much will eight of the books cost?
5. If it takes two men 6 days to paint a house, how long will it take three men painting at the same rate?

Key Words

share
divide
direct proportion
inverse proportion
reciprocal graph

Review Questions

Fractions, Decimals and Percentages

(PS) **1** **a)** Convert $\frac{13}{25}$ to a percentage. 📱 [1]

      b) Convert 0.375 to a fraction in its lowest terms. 📱 [2]

      c) Convert 36% to a fraction in its lowest terms. 📱 [2]

(FS) **2** Seema receives £5 pocket money every week. She spends $\frac{1}{2}$ of her money on magazines and $\frac{2}{5}$ on sweets. The rest she saves.

      a) How much does Seema spend on sweets? [2]

      b) How much does Seema save? [2]

(PS) **3** Work out: 📱

      a) 20% of 300 cm [2]

      b) 6% of £140 [3]

      c) 35% of 2800 g [3]

(PS) **4** A coat costing £90 is reduced by 15% in a sale. What is the sale price of the coat? 📱 [3]

(FS) **5** Karima puts £150 into a savings account. She will receive 6% simple interest each year.

      How much will she have in the bank after:

      a) 1 year? [2]

      b) 4 years? [3]

                                        Total Marks _____ / 25

(MR) **1** Write the following numbers in order of size, starting with the smallest:

      0.18 $\frac{3}{25}$ 16%   0.2 $\frac{7}{50}$ [2]

                                            Total Marks _____ / 2

Equations

(PS) **1** Solve these equations.

   a) $6x - 5 = 4x + 7$ [2]

   b) $5(x + 2) = 2(x - 1)$ [3]

   c) $3x - 1 = 4 - 2x$ [3]

   d) $\dfrac{6x - 5}{4} = 7$ [2]

2 A chocolate bar machine holds 56 bars of chocolate.

   If 29 are left, how many were sold? [2]

(FS) **3** Four builders are together paid £20 per hour plus a bonus of £150. They share the pay and each get £50.

   How many hours did they work? [3]

Total Marks _____ / 15

(PS) **1** Solve the equation $4(x - 2) - 2(3 - 2x) = 5x + 1$ [3]

> **Tip:**
> Expand the brackets first, collect like terms, then solve in the usual way.
> Remember:
> $- \times - = +$

2 Solve the equation $8 - 2a = 3(2a + 4)$ [3]

Total Marks _____ / 6

Symmetry and Enlargement

(PS) (1) Reflect shape A in the dashed mirror line.

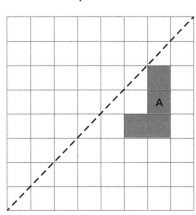

[2]

Total Marks _____ / 2

(PS) (1)

a) Rotate shape A 90° clockwise about the origin (0, 0).
Label the new shape B. [2]

b) Enlarge shape A by a scale factor 3, with centre of
enlargement (3, 4). Label the new shape C. [2]

c) Which of the shapes are congruent? [1]

d) Which of the shapes are similar? [1]

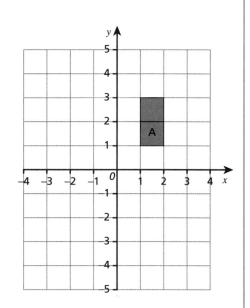

(MR) (2) A map is drawn on a scale of 1 cm : 2 km.

If a road is 13 km long in real life, how long will it be, in cm, on the map? [2]

Total Marks _____ / 8

Ratio and Proportion

PS **1** What is the ratio of black tiles to white tiles? Give the ratio in its lowest terms. [1]

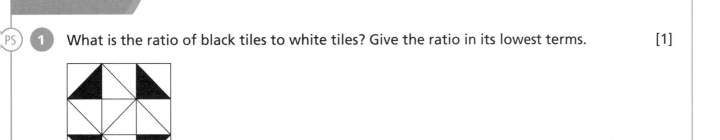

PS **2** Simplify the following ratios:

a) 8 : 24 [1]

b) 40 minutes : 1.5 hours [2]

c) £3 : 80p [2]

FS **3** Ann, Ben and Cara share £480 in the ratio 4 : 5 : 3

How much does each person get? [3]

4 A sum of money is shared in the ratio 2 : 3

If the larger share is £27, how much money is there altogether? [3]

Total Marks .... / 12

PS **1** A recipe for 6 cupcakes needs 40 g of butter and 100 g of flour.

How much butter and flour is needed to make 15 cupcakes? [2]

PS **2** Eight men can build a garage in 10 days.

Working at the same rate, how long would it take:

a) 10 men? [2]

b) 5 men? [2]

Total Marks .... / 6

Real-Life Graphs and Rates 1

Quick Recall Quiz

You must be able to:

- Read values from and draw a real-life graph
- Read and draw a conversion graph
- Solve real-life problems using graphs
- Draw a graph of exponential growth.

Graphs from the Real World

- Graphs from the real world include **conversion graphs**.
- You may also be asked to find conversions without using a graph.
- You may have to convert between:
    - pounds (£) and US dollars ($)
    - pounds (£) and euros (€)
    - pints and litres
    - mph and km/h
    - miles and kilometres
    - gallons and litres

Reading a Conversion Graph

- To convert from one unit to the other, read straight across to the line, then go straight down until you reach the other axis. To convert the other way, go up until you reach the line, then read across.

> Convert 30 miles into kilometres.
>
> Draw a line **straight up** from 30 miles until it hits the line.
>
> Go **straight across** to the kilometres axis.
>
> 30 miles is **equivalent** to 48 km.

Drawing a Conversion Graph

- To find the points you need to plot, work out a number of equivalent values. Join the plotted points with a straight line.

> Jack's company pays him 80 pence for each mile he travels. Use the information to draw a graph of his pay.
>
> | Distance in miles | 0 | 10 | 20 | 30 |
> |---|---|---|---|---|
> | Amount | 0 | £8 | £16 | £24 |
>
>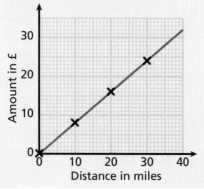

Work out how much Jack will be paid for different journeys.

Use your table to plot at least three points, and join them with a straight line.

Key Point

Extend your line to the edge of the graph grid.

The graph can now be read to find the pay for different journeys.

Solving Real-Life Problems Using Graphs

- You can use graphs to solve problems set in real-life situations.

The graph shows the charge to hire a minibus. There is a fixed charge for any distance up to 2 miles of £5 and then the charge is £3 per extra mile.

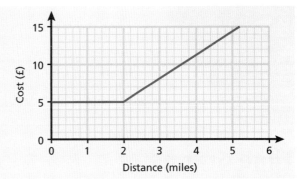

- Exponential growth means that the rate of growth is slow at the start and increases rapidly.

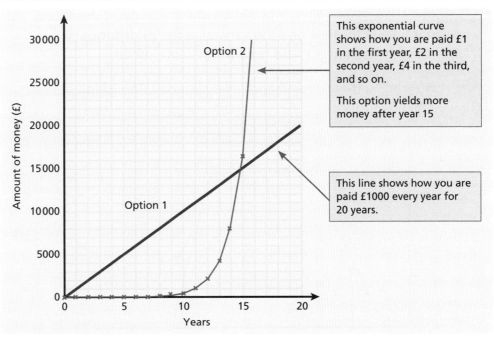

This exponential curve shows how you are paid £1 in the first year, £2 in the second year, £4 in the third, and so on.

This option yields more money after year 15

This line shows how you are paid £1000 every year for 20 years.

This **exponential graph** shows the number of infections of a disease doubling each day. How does the rate of infections change over four days?

On day 0 there are 2000 infections; day 1 has 4000; day 2 has 8000; day 3 has 16000; and day 4 has 32000
So over four days, infections have risen from 2000 to 32000

This is an increase by a factor of 16 (32000 ÷ 2000)

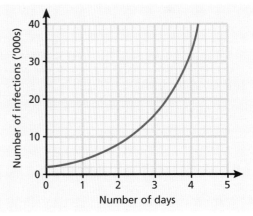

Quick Test

Use the conversion graph on page 102 for questions 1 and 2
1. How many km are equivalent to: a) 25 miles?   b) 10 miles?
2. How many miles are equivalent to: a) 30 km?   b) 40 km?
3. Sharon charges £1 for the use of her taxi and 50p per mile after that. Work out the cost of a journey that is:
   a) 4 miles   b) 6 miles
   c) Use the information to draw a graph of her charges.

Key Words

conversion graph
exponential graph

Real-Life Graphs and Rates 2

Quick Recall Quiz

You must be able to:

- Read a distance–time graph
- Work out speed, distance and time
- Work out unit prices
- Work out density, mass and volume.

Time Graphs

- Distance–time graphs give information about journeys. Use the horizontal scale for time and the vertical scale for **distance**.
- Distance–time graphs are also used to calculate **speed**.

Amanda cycles to the gym and back every Sunday. The graph below shows Amanda's journey.

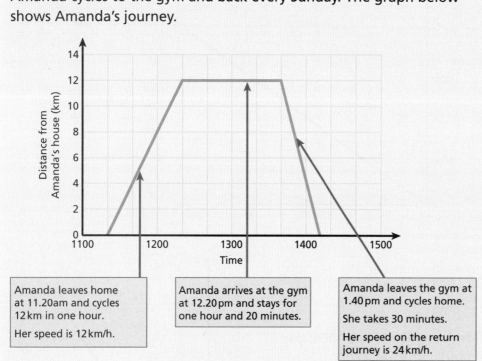

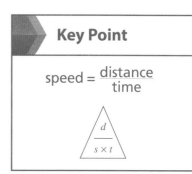

Amanda leaves home at 11.20am and cycles 12km in one hour.

Her speed is 12km/h.

Amanda arrives at the gym at 12.20pm and stays for one hour and 20 minutes.

Amanda leaves the gym at 1.40pm and cycles home.

She takes 30 minutes.

Her speed on the return journey is 24km/h.

Travelling at a Constant Speed

- When the speed you are travelling at does not change, it remains **constant**.
- You can work out speed, distance or time using a formula triangle.

A car travels 120 miles at 40 miles per hour.

How long does the journey take?

time = distance ÷ speed

time = 120 ÷ 40 = 3 hours

Key Point

$$speed = \frac{distance}{time}$$

Cover up what you are trying to find.

A plane takes $2\frac{1}{2}$ hours to travel 750 miles. What is the speed of the plane?

speed = distance ÷ time

speed = 750 ÷ 2.5

= 300 mph

Cover up what you are trying to find.

Unit Pricing

- Unit pricing involves using what 'one' is to work out other amounts.

If £1 = $1.75, how much would a pair of jeans cost in $ if they were £60?

$$60 \times 1.75 = \$105$$

How much would a TV costing $525 be in £?

$$525 \div 1.75 = £300$$

Multiply to get the dollars; divide to get the pounds.

A 250-gram bag of pasta costs £1.25

a) Work out the cost per gram.
b) Work out the number of grams bought for 1p.

a) Cost per gram = 125p ÷ 250 = 0.5p
b) Number of grams bought for 1p = 2 grams

Density

- You can work out **density**, mass and volume using a formula triangle similar to speed.

density = mass ÷ volume

Cover up what you are trying to find.

Find the density of an object that has a mass of 60 g and a volume of 25 cm³

density = mass ÷ volume

= 60 ÷ 25

= 2.4 g/cm³

> **Key Point**
>
> Remember to use the correct units.
>
> Volume: **cm³** and **m³**
>
> Mass: **g** and **kg**
>
> Density: **g/cm³** and **kg/m³**

> **Quick Test**
>
> 1. What is 90 minutes in hours?
> 2. Stuart drives 180 km in 2 hours 15 minutes. Work out his average speed.
> 3. John travelled 30 km in $1\frac{1}{2}$ hours. Kamala travelled 42 km in 2 hours. Who had the greater average speed?
> 4. If £1 = €1.20, what would £200 be worth in €?
> 5. What is the mass of 250 ml of water with density of 1 g/cm³? 1000 cm³ = 1 litre

> **Key Words**
>
> distance
> speed
> density

Right-Angled Triangles 1

You must be able to:

- Label right-angled triangles correctly
- Understand Pythagoras' Theorem
- Find the length of the longest side
- Find the length of a shorter side.

Pythagoras' Theorem

- Remember the formula for **Pythagoras' Theorem**: $a^2 + b^2 = c^2$

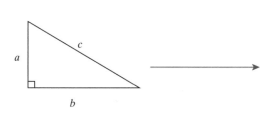

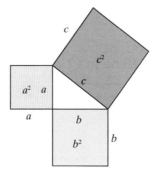

> ### Key Point
>
> The longest side, the hypotenuse, is called c and is opposite the right angle.
>
> The two shorter sides are called a and b. The order is not important.

Finding the Longest Side

- To find the longest side (**hypotenuse**), **add** the squares. Then take the **square root** of your answer.

Find the length of y. Give your answer to 1 decimal place.

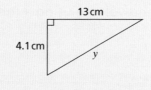

First label the sides a, b and c.

Now use the formula:

$$a^2 + b^2 = c^2$$

$$4.1^2 + 13^2 = y^2$$

$$16.81 + 169 = y^2$$

$$185.81 = y^2$$

$$y = \sqrt{185.81}$$

$$= 13.6\,\text{cm} \ (1\ \text{d.p.})$$

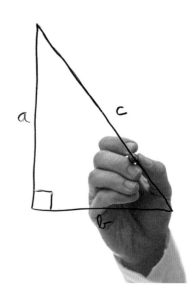

Finding a Shorter Side

- To find a shorter side, **subtract** the squares. Then take the square root of your answer.

Find the length of y. Give your answer to 1 decimal place.

 Label the sides a, b and c.

Now use the formula:

$a^2 + b^2 = c^2$

$7^2 + y^2 = 14^2$

$49 + y^2 = 196$

$y^2 = 196 - 49 = 147$

$y = \sqrt{147} = 12.1\,\text{cm}$ (1 d.p.)

Find the length of y. Give your answer to 1 decimal place.

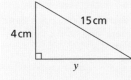 Label the sides a, b and c.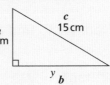

Now use the formula:

$a^2 + b^2 = c^2$

$4^2 + y^2 = 15^2$

$16 + y^2 = 225$

$y^2 = 225 - 16 = 209$

$y = \sqrt{209} = 14.5\,\text{cm}$ (1 d.p.)

Key Point

You will always $\sqrt{}$ at the end.

Quick Test

1. Work out:
   a) 3.2^2
   b) 15.65^2
2. Work out:
   a) $\sqrt{4900}$
   b) $\sqrt{39.69}$
3. Work out the longest side of a right-angled triangle if the shorter sides are 5 cm and 2.2 cm.
4. Work out the Shorter side of a right-angled triangle if the longest side is 12 cm and the other shorter side is 9 cm.

Key Words

Pythagoras' Theorem
hypotenuse
square

Right-Angled Triangles 2

Geometry and Measures

You must be able to:

- Remember the three ratios
- Work out the size of an angle
- Work out the length of an unknown side.

Side Ratios

- Label the sides of the triangle in relation to the angle that is marked.

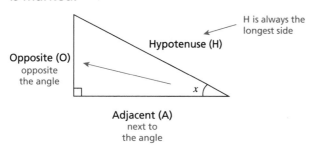

H is always the longest side

Hypotenuse (H)

Opposite (O) opposite the angle

x

Adjacent (A) next to the angle

- There are three ratios: **sin**, **cos** and **tan**. Try to find a way of remembering these:

Sin is short for 'sine', cos is short for 'cosine' and 'tan' is short for 'tangent'.

$$\sin x = \frac{O}{H} \qquad \cos x = \frac{A}{H} \qquad \tan x = \frac{O}{A}$$

- You can use the formula triangles:

- You can use a rhyme:
  Some **O**ld **H**orses **C**an **A**lways **H**ear **T**heir **O**wners **A**pproach

Use your calculator to work out the ratios for these angles:

a) $\sin 60° = 0.8660$
b) $\cos 45° = 0.7071$
c) $\tan 87° = 19.0811$

> **Key Point**
>
> Ensure your calculator is in '**degree**' mode.

Use your calculator to work out the angles for these ratios:

a) $\sin x = 0.5$ $x = \sin^{-1} 0.5 = 30°$

b) $\cos x = \frac{3}{5}$ $x = \cos^{-1} (3 \div 5) = 53.1°$

c) $\tan x = 2.9$ $x = \tan^{-1} 2.9 = 71°$

Finding Angles in Right-Angled Triangles

- You need to know two sides of the triangle to find an angle.

Find the size of angle x. Give your answer to 1 decimal place.

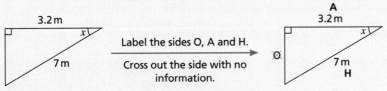

Label the sides O, A and H.

Cross out the side with no information.

As **A** and **H** are known, we use the cos ratio.

$\cos x = \dfrac{A}{H}$ so $\cos x = \dfrac{3.2}{7}$

$x = \cos^{-1}\left(\dfrac{3.2}{7}\right)$

$x = 62.8°$

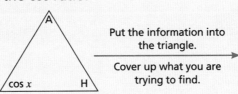

Put the information into the triangle.

Cover up what you are trying to find.

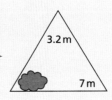

Finding the Length of a Side

- You need to know the length of one side and an angle to find the length of another side.

Find the length of the side labelled y. Give your answer to 1 decimal place.

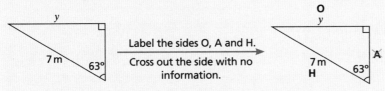

Label the sides O, A and H.

Cross out the side with no information.

As **O** is to be found and **H** is known, we use the sine ratio.

$\sin x = \dfrac{O}{H}$ so $\sin 63° = \dfrac{y}{7}$

$7 \times \sin 63° = y$

$y = 6.2\,\text{m}$

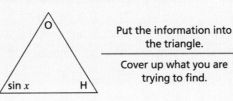

Put the information into the triangle.

Cover up what you are trying to find.

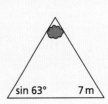

> **Quick Test**
>
> 1. Use your calculator to work out the ratios for these angles.
>   a) $\sin 20°$ b) $\cos 30°$ c) $\tan 45°$
> 2. Use your calculator to work out the angles for these ratios.
>   a) $\sin x = 0.8337$ b) $\cos x = \dfrac{4}{7}$ c) $\tan x = 32$

> **Key Words**
>
> sin
> cos
> tan

Review Questions

Symmetry and Enlargement

(PS) **1** Reflect shape A in the dashed mirror line.

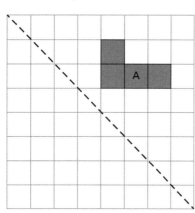

[2]

Total Marks _____ / 2

(PS) **1** **a)** Rotate shape A 180° about the point (2, 2).
Label the new shape B. [2]

b) Enlarge shape A by a scale factor of 2, with centre
of enlargement (2, 3). Label the new shape C. [2]

c) What is the area of shape A? [1]

d) What is the area of the shape C? [1]

e) What is the ratio of the areas A : C? [1]

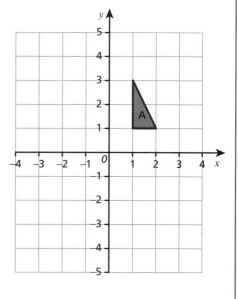

(MR) **2** The diagrams below represent two photographs.

Find the width of the enlarged photograph. [2]

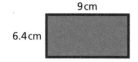

Total Marks _____ / 9

Ratio and Proportion

(PS) **1** Simplify the following ratios.

    a) 16 : 24   [1]

    b) 25p : £2   [2]

    c) 0.75 km : 200 m   [2]

2 Complete the equivalent ratios.

    a) 8 : 3 = ? : 15   [1]

    b) 7 : ? = 63 : 108   [1]

    c) 4 : 5 = 6 : ?   [1]

3 Share 40 pens in the ratio 3 : 5   [2]

(FS) **4** A sum of money is shared in the ratio 1 : 4 : 3

    If the largest share is £120, how much money is there altogether?   [3]

Total Marks _____ / 13

(PS) **1** A machine can produce 1140 plastic cups in 8 hours. At the same rate, how many plastic cups can be made in:

    a) 10 hours?   [2]

    b) 12 hours?   [2]

(PS) **2** If it takes two men 3 days to paint a room, how long would it take three men to paint the same room?   [2]

(MR) **3** Twelve bags of oats will be enough for three donkeys for eight days.

    How long will 10 bags last four donkeys at the same rate?   [3]

Total Marks _____ / 9

Real-Life Graphs and Rates

FS 1 The table shows the prize money for the winners of professional tennis tournaments in Australia and France in one particular year.

Country	Australia	France
Money	1 000 000 Australian dollars (£1 = 2.70 Australian dollars)	780 000 euros (£1 = 1.54 euros)

Which country paid more money? You must show your working. [2]

PS 2 The graph shows the flight details of an aeroplane travelling from London to Madrid via Brussels.

What is the aeroplane's average speed from London to Brussels?

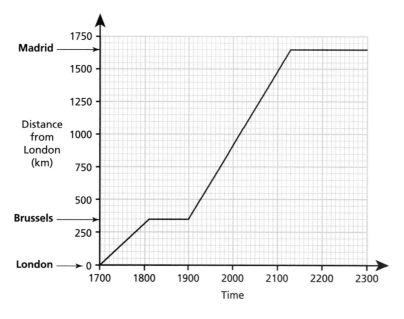

[2]

Total Marks _____ / 4

PS 1 **a)** Calculate the density of a piece of metal that has a mass of 2000 kg and a volume of 0.5 m³ [2]

b) Calculate the volume of the same type of metal that has a mass of 5000 kg. [2]

Total Marks _____ / 4

Right-Angled Triangles

(PS) **1** Use Pythagoras' Theorem to work out the following:

    a) The length of BC **b)** The length of AC

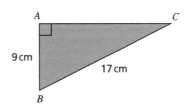

[4]

(PS) **2** **a)** Work out the value of p.   **b)** Work out the size of angle y.

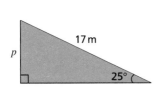 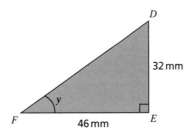

[4]

Total Marks _____ / 8

(PS) **1** Work out the size of angle x.

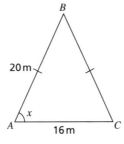

[3]

(MR) **2** State whether or not each triangle is right-angled if the sides measure:

    a) 7 cm, 12 cm, 18 cm

    b) 10 cm, 24 cm, 26 cm

    c) 11 cm, 19 cm, 21 cm

    d) 7 cm, 24 cm, 25 cm

[4]

Total Marks _____ / 7

Review Questions

Real-Life Graphs and Rates

(FS) **1** Use £1 = US$1.75 to work out how much:

    a) US$200 is in £ [2]

    b) £200 is in US$ [2]

(PS) **2** A coach travels 300 miles non-stop at an average speed of 40 mph.

    a) For how many hours does the coach travel? [2]

    b) At the same speed, how far will the coach travel in four hours? [2]

Total Marks _____ / 8

(PS) **1** At time $t = 0$, one bacteria is placed in a dish in a laboratory. The number of bacteria doubles every 10 minutes.

    a) Draw a graph to show the growth of bacteria over 100 minutes. [3]

    b) Use your graph to estimate the time taken to grow 300 bacteria. [1]

Time (t minutes)	No. of bacteria
0	1
10	
20	
30	
40	
50	
60	
70	
80	
90	
100	

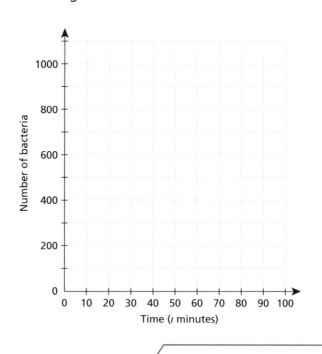

Total Marks _____ / 4

Right-Angled Triangles

1 **a)** Work out the size of angle x. **b)** Work out the length of x.

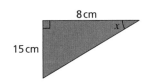

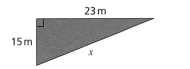

[4]

2 **a)** Work out the length of y. **b)** Work out the length of AB.

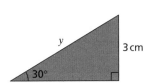

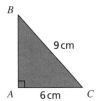

[4]

3 A wire 18 m long runs from the top of a pole to the ground as shown in the diagram. The wire makes an angle of 35° with the ground.

Calculate the height of the pole. Give your answer to a suitable degree of accuracy.

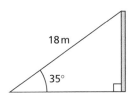

[2]

Total Marks _____ / 10

1 **a)** Work out the vertical height of triangle ABC.

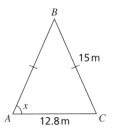

[2]

b) Work out the size of angle x.

[2]

Total Marks _____ / 4

Mixed Test-Style Questions

No Calculator Allowed 🖩

1 Work out both the surface area and volume of these cuboids.

a)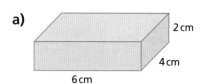

Surface area = ... cm²

Volume = ... cm³

b)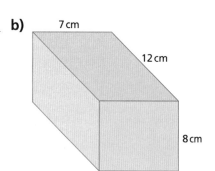

Surface area = ... cm²

Volume = ... cm³

4 marks

2 Solve the following, giving your answers in the simplest form.

a) $4\frac{1}{2} + 2\frac{1}{3}$

b) $5\frac{2}{3} + 8\frac{1}{4}$

c) $9\frac{1}{6} - 2\frac{3}{8}$

d) $12\frac{1}{2} - 14\frac{5}{6}$

4 marks

3 **a)** Plot the following coordinates onto the grid below: (3, 5) (1, 5) (2, 7)

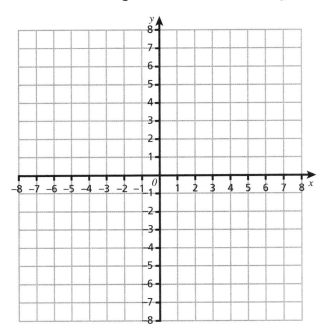

b) The three points are vertices of a rhombus.

Work out the coordinates of the fourth vertex.

2 marks

4 **a)** Complete the table of values for the equation of $y = -2x + 3$

x	−2	−1	0	1	2	3
y						

b) Plot the coordinates on the grid below and join them with a line.

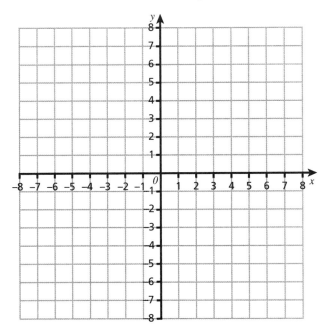

4 marks

TOTAL

14

Mixed Test-Style Questions

5   Calculate angles x and y.

a)

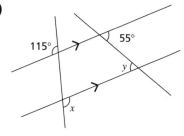

$x =$ .... °

$y =$ .... °

b)

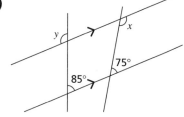

$x =$ .... °

$y =$ .... °

4 marks

---

6   **a)**   Simplify the following expressions:

     i) $3x - 2y + x + 6y$

     ii) $4g + 5 - g - 4$

   b)   Expand and simplify the following expressions:

     i) $4(x - 5)$

     ii) $4x(x + 4)$

   c)   Factorise completely the following expressions:

     i) $6x - 12$

     ii) $4x^2 - 8x$

6 marks

7 The rectangle and trapezium below have the same area.

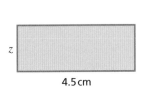
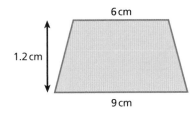

Work out the length of z. Show your working.

3 marks

8 **a)** Two numbers have a sum of −5 and a product of 6

Work out the two numbers.

b) Two different numbers have a sum of 7 and a product of −8

Work out the two numbers.

2 marks

TOTAL

15

9 The shape below is made from four congruent triangles like the one on the right.

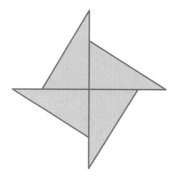

a) Write an expression in terms of a, b and c for the perimeter of the shape.

b) Given that $a = 3$, $b = 4$ and $c = 5$, find the value of the perimeter.

4 marks

10 A certain plant grows by 10% of its height each day. At 8 am on Monday the plant was 400 mm high.

How tall was it:

a) at 8 am on Tuesday?

b) at 8 am on Wednesday?

2 marks

11 **a)** Solve the equation $3y - 2 = 13$

b) Solve the equation $3 - \frac{x}{4} = -5$

4 marks

12

a) Reflect shape A in the y-axis.

b) Enlarge shape A by a scale factor of 2 from the point (3, 4)

c) Rotate shape A 180° from (0, 0)

3 marks

TOTAL

13

Mixed Test-Style Questions

Calculator Allowed

1 Sam was sitting on the dock of the bay watching boats for an hour. He collected the following information:

Type	Frequency	Probability
Tug boat	12	
Ferry boat	2	
Sail boat	16	
Speed boat	10	

a) Complete the probability column in the table, giving your answers as fractions.

b) If Sam saw 75 boats, estimate how many of them would be sail boats.

5 marks

2 Work out the surface area and volume of these cylinders.

a) radius = 4 cm

9 cm

Surface area = _____ cm²

Volume = _____ cm³

b) diameter = 10 cm

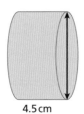

4.5 cm

Surface area = _____ cm²

Volume = _____ cm³

8 marks

3 The diagram shows a circle inside a square of side length 4 cm.

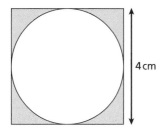

4 cm

Find the total area of the shaded regions.

_____ cm² [ ]

3 marks

4 Barry is planning to buy a car. He visits two garages which have the following payment options:

Mike's Motors	Carol's Cars
£500 deposit	£600 deposit
36 monthly payments of £150	12 monthly payments of £50
£150 administration fee	24 monthly payments of £200

Which garage should Barry buy his car from in order to get the cheaper deal? Show working to justify your answer.

[ ]

3 marks

TOTAL

[ ]

19

5 The nth term of a sequence is given by the expression $\dfrac{(n+3)(n+4)}{2}$

a) Write down an expression in terms of n for the $(n + 1)$th term.

b) Use the two expressions to prove that the sum of two consecutive terms in the sequence is a square number.

4 marks

6 A wire 15 m long runs from the top of a pole to the ground as shown in the diagram. The wire makes an angle of 45° with the ground.

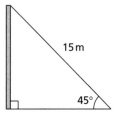

Calculate the height of the pole. Give your answer to a suitable degree of accuracy.

2 marks

7 Donny the magician claims to be able to read minds.

His friend Lewis asks him to prove his claim.

Donny tells Lewis to think of a number and to follow the instructions given below. Donny will know the answer.

The instructions are as follows:

Think of a number, multiply it by 2, add 10, divide by 2, and subtract the number you first thought of.

Donny tells Lewis he got the answer 5 and he is right.

Complete the table below to show why Donny's trick worked.

Instruction	Mathematical expression
Think of a number	n
Multiply by 2	
Add 10	
Divide by 2	
Subtract the number you thought of	

2 marks

8 Solve $\frac{2x}{3} = 36$

2 marks

TOTAL

10

9 A recipe for 12 cupcakes needs 80 g of butter and 200 g of flour.

How much butter and flour are needed to make:

a) 24 cupcakes?

....g of butter

....g of flour

b) 30 cupcakes?

....g of butter

....g of flour

4 marks

10 The mean of 7 numbers is 11

I add another number and the mean is now 12

What number did I add?

2 marks

11 The data gives the waiting time in minutes of 15 patients in a surgery:

49	23	34	10	28	28	25	45
39	35	15	14	48	10	20	

Work out the median waiting time.

2 marks

12 Thirty children were asked if they liked cycling or swimming.

Set A is those who like cycling.

Set B is those who like swimming.

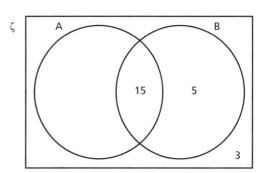

a) Complete the Venn diagram.

b) How many children do not like either cycling or swimming?

c) Find P(A ∩ B)

3 marks

TOTAL

11

Answers

Page 6

1. 1996 [1] because 2000 – 1996 = 4 but 2007 – 2000 = 7 [1]
2.

4	7	6	245	[2]
– 2	3	1	**OR** Method mark for valid attempt to subtract	[1]
2	4	5		

3. 5000, 46 000, 458 000, 46 000
   All 4 correct [2] Any 3 correct [1]
4. $\frac{27}{50}$, 55%, 0.56, 0.6, 0.63
   All 5 correct [3] **OR** 3 out of 5 correct [2] **OR** 55% = 0.55 and $\frac{27}{50}$ = 0.54 [1]
5. Ahmed = 2 × 22 = 44 years old   [2]
   OR Rebecca = 25 – 3 = 22 years old seen   [1]
6. 15 878   [3]
   OR 12 000 + 1800 + 210 + 1600 + 240 + 28   [2]
   OR Valid attempt at multiplication with one numerical error [1]
7. 52   [2]
   OR Valid attempt at division with one numerical error   [1]
8.

   Both correct [2]; One correct [1]

Page 7

1. 200 ml   [3]
   OR 1000 ml and 800 ml seen   [2]
   OR 1000 ml seen   [1]
2. a) 16 **OR** 36   [1]
   b) 13 or 17 or 31 or 37 or 53 or 61 or 67 or 71 or 73   [1]
   c) 36   [1]
   d) 15   [1]
3. 9 cm [2] **OR** 27 ÷ 3 seen [1]

   > An equilateral triangle has three equal sides.

4. 33   [2]
   OR 11 seen   [1]
5. 25°   [3]
   OR 65 seen   [2]
   OR 130 seen   [1]

   > Angles in a triangle add up to 180°. Base angles in an isosceles triangle are equal. A right angle is 90°.

6. T = 200 [1] (S = T + 100, 2T + 100 = 500) S = 300 [1]

Page 9 Quick Test
1. 35
2. 226 635
3. 163
4. 64

Page 11 Quick Test
1. a) 49 b) 64
2. a) 7 b) 3
3. $2^3 \times 5$
4. 252
5. 8

Page 13 Quick Test
1. ÷2 **OR** × $\frac{1}{2}$
2. 5, 9, 13, 17, 21... (rule +4)
3. –3
4. 4, 8, 16, 32, 64... (rule ×2)

Page 15 Quick Test
1. 8, 13, 18, 23, 28
2. 4, 19, 44, 79, 124
3. a) 24 – 4n b) –176
4. Position-to-term rule

Page 16

1. Jessa is right as using BIDMAS multiply is first [2] **OR** Using BIDMAS the multiply is first [1]
2. a) £4248 [3] **OR** 3000 + 500 + 40 + 600 +100 + 8 [2] **OR** Valid attempt at multiplication with one numerical error [1]
   b) 354 ÷ 52 = 6.8 [2] **OR** Valid attempt at division with one numerical error [1] 7 coaches needed [1]
   c) 10 spare seats [2] **OR** 364 seen (52 × 7 = 364) [1]
3. Harry's because 3 × 6 = £18 but 5 × 4 = £20 [3] **OR** £18 and £20 seen but no conclusion [2] **OR** 15 ÷ 3 and 15 ÷ 5 seen [1]
4. –3   [1]

1. p = 3 [1], q = 3 [1], r = 5 [1] **OR**
   Valid attempt at prime factorisation seen   [1]
2. xy [1]

Page 17

1. a) 3n + 1 [3] **OR** 3n [2] **OR** +3 seen as term-to-term rule [1]
   b) 181 [1]
2. 5, 9, 13, 17, 21 = arithmetic. 2, 8, 18, 32, 50 = quadratic. 8, 17, 32, 53, 80 = quadratic. All three correct [2] **OR** one correct [1]
3. a) 8^2 = 64 and 9^2 = 81 so 79 is between 8 and 9 **OR** attempt to find any two square numbers each side of 79 [2]
   b) 8.89   [1]

1. 2, 7, 12, 17, 22 = 5n – 3
   3, 9, 27, 81 = 3^n
   6, 21, 46, 81, 126 = 5n^2 + 1
   4, 3, 2, 1, 0 = 5 – n
   All four correct [2]; Two correct [1]
2. 25 units [2] **OR** attempt to divide by 2 once or more [1]

Page 19 Quick Test
1. 24 cm
2. 27 cm²
3. 6 cm²
4. Area 19 cm²
   Perimeter 24 cm

Page 21 Quick Test
1. 16 cm²
2. 13 cm²
3. Circumference = 18.8 cm (1 d.p.)
   Area = 28.3 cm² (1 d.p.)
4. Area = 5.6 cm² (1 d.p.)

Page 23 Quick Test
1. a) Red = 9; Blue = 7; Green = 8; Yellow = 6; Other = 7
   b) 37
2. a) 6 and 3
   b) Mean = 9.2 (1 d.p.); Median = 6; Range = 37
   c) Median as there is an outlier

Page 25 Quick Test
1. a)

	Football	Rugby	Total
Women	16	9	25
Men	20	10	30
Total	36	19	55

   b) 36   c) 9   d) 55

Page 26

1. 1248 **[1]** (24 is half of 48) 26 **[1]** (26 is half of 52) 48 **[1]**
2. 15 and 12 **[2] OR** either 15 or 12 seen **[1]**
3. 4 packs of sausages and 3 packs of rolls **[3] OR** 24 seen **[2]**
   OR Valid attempt to find LCM of 6 and 8 seen **[1]**

1. a) $x = 2$ **[1]** $y = 3$ **[1]**
   b) $z = 5$ **[1]**

   > Remember $x^m \times x^n = x^{m+n}$

2. Can be written as $(5x)^2$ **[1]**
3. –1 and –4 **[1]**
4. No. Any counter example, e.g. $1 + -8 = -7$ **[2]**

Page 27

1. a) 44 **[1]**

   > For 20th term, $n = 20$

   b) 204 **[1]**
2. $2n - 1$ **[2] OR** $2n$ seen **[1]**
3. a) 14 **[1]**
   b) 9 weeks **[1]**
   c) $3n + 5$ **[1]**

1. Cindy is right **[1]** BIDMAS states indices first so $4 \times 100 + 2 = 402$ **[1]**
2. 28 days **[2] OR** $\frac{540}{20}$ seen **[1]**

Page 28

1. $x = 8\,\text{cm}$ **[1]** $y = 6.8\,\text{cm}$ **[1]**

   > Area of a rectangle $= l \times w$

2. a) 192 **[3] OR** 16×12 seen **[2] OR** 120 000 and 625 seen **[1]**

   > As each tile is 25 cm, 16 will fit along one side and 12 along the other.

   b) £300 **[2] OR** 20 seen **[1]**

   > The nearest multiple of 10 bigger than 192 is 200

   c) 8 tiles **[1]**

1. $20\,\text{cm}^2$ **[3] OR** 10 seen as an attempt to find area of triangle **[2]**
   OR 2 and 5 seen as an attempt to find midpoints. **[1]**

Page 29

1. a) Phil 66.4 (1 d.p.) **[1]** Dave 68.9 (1 d.p.) **OR** attempt to add them up and divide by number of values **[1]**
   b) Phil 70 **[1]** Dave 175 **[1]**
   c) Dave as higher average **OR** Phil as more consistent **[2]**

1. 16.7 mm **[3] OR** 870 seen **[2] OR** 5, 15, 30, 50 seen as midpoints **[1]**

Page 31 Quick Test

1. a) 235.6   b) 5.6781
2. 100 000
3. 16.2, 16.309, 16.34, 16.705, 16.713

Page 33 Quick Test

1. 49.491
2. 17.211
3. 163
4. 150 000

Page 35 Quick Test

1. $7x + 5y + 6$
2. c^2d^2
3. 26

Page 37 Quick Test

1. $8x - 4$
2. $2x - 14y$
3. $5(x - 5)$
4. $2x^3(x^2 - 2)$

Page 38

1. a) $25\,\text{m}^2$ **[3] OR** 20 and 5 seen **[2]**
   OR only 20 seen **[1]**
   b) 4 tins **[1]** c) £48 **[2] OR** answer to b) × 12 seen **[1]**

   > This is a compound area; separate into a rectangle and triangle.

2. Any answer from 5092–5095 (depending on value of pi used) **[2]**
   OR 157 or 1.57 seen **[1]**

   > This question is about the circumference of a circle; notice the units are different; 50 cm = 0.5 m

1. a) Sector B as 8.73 is bigger than 5.65 **[3] OR** 5.65 or 8.73 seen as an attempt to find area of sector **[2] OR** 28.3 or 78.5 seen as attempt to find the area of circle **[1]**
   b) Sector B, as 13.49 is bigger than 9.77 **[3] OR** 3.77 or 3.49 seen as an attempt to find an arc length **[2] OR** 18.8 or 31.4 seen as an attempt to find the circumference of a circle **[1]**

Page 39

1. 30 **[3] OR** Two-way table seen with no more than two mistakes **[2] OR** Two-way table with no more than three mistakes. **[1]**

	Home	Hospital	Water	Total
Teen	2	10	4	16
Non-teen	18	20	6	44
Total	20	30	10	60

2. Median as data contains an outlier (14 808 much bigger than the rest of the data and there is no mode) **[1]**

1.
   > To find the total, multiply the mean by the number of pupils.

   6.2 **[3] OR** 174 seen **[2] OR** 210 seen **[1]**
2. Mode, as this is the most common size required **[2] OR** Mode **[1]**

Page 40

1. a) 57.832 **[1]**
   b) 21.98 **[1]**
   c) 74.154 **[1]**
   d) 216 **[1]**
2. 16.061, 16.94, 17.09, 17.203, 17.84 **[2]**
   [1] for any four in correct order, ignoring fifth value

1. $0.02 \to 200$, $50 \to 0.08$, $8 \to 0.5$, $20 \to 0.2$
   All 4 correct **[2] OR** 2 correct **[1]**
2. a) 6.89×10^6 **[1]**
   b) 8.766×10^{-3} **[1]**
   c) 59 890 000 is bigger as $5.989 \times 10^4 = 59\,890$ **[1]**

Page 41

1. Both of them **[1]** $2(x + y)$ expands to $2x + 2y$ or vice versa **[1]**
2. £123 **[3] OR** 48 seen **[2] OR** 120×0.4 seen **[1]**
3. $3x + 7$ **[2] OR** $8x + 2 - 5x + 5$ seen **[1]**

4. $3a(bc + 2)$ **[2] OR** $3(abc + 2a)$ or $a(3bc + 6)$ seen **[1]**

'Completely' means remove all factors.

5. 144 **[2] OR** 9 seen **[1]**

$ab = a \times b$

6. $3a - b$, $2a - b$, b
All three correct **[2]**; any one correct **[1]**

1. $x^2 + 2x - 8$ **[2] OR** $x^2 + 4x - 2x - 8$ **[1]**

Always remember to simplify.

2. $(x + 2)$ **[1]** $(x - 1)$ **[1]**

Factorising is the opposite of expanding.

3. $(x + y)^2 = (x + y)(x + y) = x^2 + 2xy + y^2$ **[2] OR** $(x + y)^2 = (x + y)(x + y)$ **[1]**

Page 43 Quick Test
1. Volume = 140 cm³
Surface area = 166 cm²
2. Volume = 440 cm³
Surface area = 358 cm²

Page 45 Quick Test
1. a) Volume = 603 cm³
Surface area = 402 cm²
b) Volume = 1800 cm³
Surface area = 1620 cm²
2. 116 cm³

Page 47 Quick Test
1. 20°
2. Sunday
3. No, Helen still used more (approx. mean 51).

Page 49 Quick Test
1. Own goal
2. Positive
3. The spinner is biased **OR** there is no yellow on the spinner.

Page 50

1. 6.765, 6.776, 7.675, 7.756, 7.765
All five correct **[2] OR** three correct **[1]**
2. a) £15.99 **[2] OR** attempt to add three costs with only one numerical mistake **[1]**
b) £4.01 **[1]**

1. a) £8.20 **[1]**
b) £8.30 **[1]**
c) 1.2% **[2] OR** $10 \div 830$ or $\frac{0.1}{83}$ seen **[1]**

Page 51

1. a) ab **[1]**
b) $2a + 2b$ or $2(a + b)$ **[1]**
c) $3a$ and $5a$ **[1]**
2. 2 **[1]** −2 **[1]**
3. $4t(2ut - u + 5)$ **[2] OR** $4(2ut^2 - ut + 5t)$ **[1] OR** $t(8ut - 4u + 20)$ **[1]**
4. Yes as $4n$ can be written as $2(2n)$ **[1]**

Integer means whole number.

1. a) $x^2 - y^2$ **[1]**
b) 800 seen with 201 and 199 substituted into the brackets **[2] OR** 400 and 2 seen **[1]**

2. Yes **[1]** with 190 to nearest cm, largest value 190.5. 200 cm to nearest 10 cm, smallest 195 cm. **[1]** 195 is bigger than 190.5 **[1] OR** Yes **[1]** with 190.5 or 195 seen **[1] OR** Just 190.5 or 195 seen. **[1]**

Page 52
1. Surface area = 117 cm² **[1]**
Volume = 81 cm³ **[1]**
2. $942 \div 5^2 \div \pi$ **[1]** = 12.0 cm **[1]**
3. $1385 \div 7^2 \div \pi$ **[1]** = 9.00 cm **[1]**

1. Surface area = 672 cm² **[1]**
Volume = 960 cm³ **[1]**
2. a) $2 \times 2 \times 10 = 40$ **[1]** $14 - 2 = 12$ **[1]** $12 \times 2 \times 2 = 48$
$40 + 48 = 88$ cm³ **[1]**
b) $14 \times 7 \times 4 = 392$ **[1]** $5 \times 2 \times 4 = 40$ **[1]**
$392 - 40 = 352$ cm³ **[1]**

Page 53

1. Example answer: Fuel remaining **[1]** against distance travelled **[1]**
2. Question should have a time frame (e.g. How many more times do you go shopping during the Christmas period than other times of the year?). **[2]**
Response boxes should cover all outcomes and not overlap (e.g. about the same; 1–5 times more; 6–10 times more; more than 10 times more). **[2]**
3. Scatter, frequency polygon, pie chart, bar chart, line graph, histogram, etc. Four examples **[2]**
Choose a graph with a reason, for instance pie chart as it shows the percentage of time spent. **[2]**
4. Paper 2 was more difficult **[1]**
Valid reason, e.g. More students scored the lower marks on paper 2 **OR** More students scored high marks on paper 1 **[1]**

1. Hypothesis – The dice is biased **OR** has no number 8 **[1]**. Test roll the dice recording outcomes a large number of times **[1]**. If the outcomes are fairly split then the dice is not biased **[1]**.

Page 55 Quick Test
1. e.g. $\frac{4}{6}$ $\frac{6}{9}$ $\frac{8}{12}$ $\frac{10}{15}$ $\frac{12}{18}$
2. $\frac{64}{77}$
3. $\frac{29}{72}$
4. $\frac{15}{52}$
5. $\frac{81}{100}$

Page 57 Quick Test
1. $\frac{1}{3}$
2. $\frac{49}{36} = 1\frac{13}{36}$
3. $\frac{28}{5} = 5\frac{3}{5}$
4. $\frac{37}{4} = 9\frac{1}{4}$
5. $2\frac{17}{90}$
6. $3\frac{19}{20}$
7. $2\frac{8}{9}$

Page 59 Quick Test
1.

x	−2	−1	0	1	2	3
y	−11	−8	−5	−2	1	4

A line graph should be accurately plotted using the values in the table.
2. $y = 5x + 3$

Page 61 Quick Test

1. a) Gradient = 3   Intercept = 5
   b) Gradient = 6   Intercept = –7
   c) Gradient = –3   Intercept = 2

2.

x	–3	–2	–1	0	1	2	3
y	4	2	2	4	8	14	22

Pages 62–63 Review Questions

Page 62

1. Surface area = $2(14 \times 2) + 2(6 \times 2) + 2(14 \times 6)$ **[1]** = 248 cm² **[1]**
   Volume = $14 \times 2 \times 6$ **[1]** = 168 cm³ **[1]**
2. Radius = $\frac{11}{2}$ = 5.5 m **[1]** $5.5^2 \times \pi$ = 95.03 **[1]**
   95.03×2.2 = 209 m³ **[1]**
3. $\sqrt[3]{512}$ **[1]** = 800 cm **[1]**
4. No Parveen does not have enough. **[1]**
   Surface area = $2(15 \times 10 + 15 \times 20 + 10 \times 20)$ **[1]** = 1300 cm² **[1]**

---

1. Volume = $(8 \times 12 \times 6) + (3 \times 8 \times 12 \times \frac{1}{2})$ **[1]** = 576 + 144 = 720 m³ **[1]**
   Surface area = $2(8 \times 6) + 2(12 \times 6) + (12 \times 8) + 2(12 \times 5) + 2(3 \times 8 \times \frac{1}{2})$ **[1]** = 96 + 144 + 96 + 120 + 24 = 480 m² **[1]**

Page 63

1.

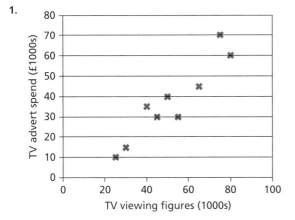

Correct axes labels **[1]**. Points plotted correctly **[1]**.
It has a positive correlation **[1]**; as advertising spend increases, viewing figures also increase. **[1]**

2. a) Example answers: 'a lot' is too vague – the quantity should be specific **[1]**; how much junk food in a period of time could be asked **[1]**
   b) Example answers: Needs to cover all possible options, e.g. 0 and greater than 4 **[1]**; quantity of fruit could be more specific, e.g. portion size **[1]**

---

1. Example answers: Laura spends the greatest proportion of her time in the garden whereas Jules spends it watching TV. Jules doesn't walk at all and goes to the gym more than Laura, etc.   Three comparisons **[3]**
2. Katya could hypothesise that the spinner is biased. **[1]** She would test by spinning it many times **[1]** and finding the experimental probability for each colour. **[1]**

Pages 64–65 Practice Questions

Page 64

1. a) $\frac{21}{20} = 1\frac{1}{20}$ **[1]**
   b) $\frac{41}{40} = 1\frac{1}{40}$ **[1]**
   c) $\frac{29}{20} = 1\frac{9}{20}$ **[1]**
   d) $\frac{5}{8}$ **[1]**

e) $\frac{3}{10}$ **[1]**
f) $\frac{19}{36}$ **[1]**

2. a) $\frac{2}{24} = \frac{1}{12}$ **[1]**
   b) $\frac{40}{54} = \frac{20}{27}$ **[1]**
   c) $\frac{3}{20}$ **[1]**
3. a) $\frac{3}{16}$ **[1]**
   b) $\frac{9}{48}$ **[1]** $= \frac{3}{16}$ **[1]**
   c) $\frac{21}{12} + \frac{1}{2} = \frac{9}{4}$ **[1]** $= 2\frac{1}{4}$ **[1]**

---

1. a) $\frac{35}{8} + \frac{11}{5} = \frac{263}{40}$ **[1]** $= 6\frac{23}{40}$ **[1]**
   b) $\frac{18}{5} + \frac{21}{9} + \frac{7}{2} = \frac{283}{30}$ **[1]** $= 9\frac{13}{30}$ **[1]**
   c) $\frac{29}{4} - \frac{30}{11} = \frac{199}{44}$ **[1]** $= 4\frac{23}{44}$ **[1]**
   d) $\frac{11}{5} - \frac{10}{7}$ **[1]** $= \frac{27}{35}$ **[1]**
   e) $\frac{3}{2} \times \frac{14}{3}$ **[1]** $= 7$ **[1]**
   f) $\frac{16}{5} \div \frac{7}{5} = \frac{16}{5} \times \frac{5}{7}$ **[1]** $= \frac{16}{7}$ **[1]** $= 2\frac{2}{7}$ **[1]**

Page 65

1. (–1, 4) **[1]** and (–3, 2) **[1]**

2.

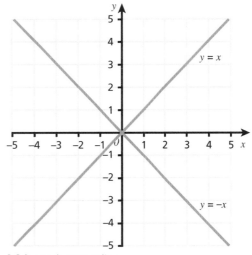

[2]
[1] for each correct line.

3.

x	–1	0	1	2	3
y	7	4	1	–2	–5

[2]

[1] for at least three correct values.

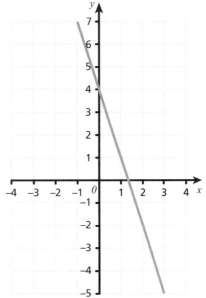

[2]
[1] for at least three correct values plotted.

1. Gradient = –3 **[1]** and *y*-intercept is (0, 4) or +4 **[1]**
2.

x	–3	–2	–1	0	1	2	3
y	16	8	2	–2	–4	–4	–2

All correct **[3]**; At least five correct **[2]**; At least three correct **[1]**

Pages 66–73 **Revise Questions**

Page 67 Quick Test
1. Student's own drawings
2. **a)** 76° **b)** 56° **c)** 65°

Page 69 Quick Test
1. **a)** 55° **b)** 112° **c)** 126°
2. Equilateral triangle, square or hexagon

Page 71 Quick Test
1. Likely
2.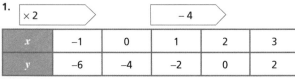

3. **a)** $\frac{1}{3}$ **b)** $\frac{2}{3}$
4. 0.228

Page 73 Quick Test
1. **a)** 0.47 **b)** 0.53 **c)** 9.5 so 9 or 10

Pages 74–75 **Review Questions**

Page 74

1. **a)** $\frac{4}{10} + \frac{1}{10} = \frac{5}{10} = \frac{1}{2}$ **[1]**
   b) $\frac{7}{12} + \frac{3}{12} = \frac{10}{12} = \frac{5}{6}$ **[1]**
   c) $\frac{5}{30} + \frac{6}{30} = \frac{11}{30}$ **[1]**
   d) $\frac{20}{70} + \frac{21}{70} = \frac{41}{70}$ **[1]**
   e) $\frac{8}{9} - \frac{3}{9} = \frac{5}{9}$ **[1]**
   f) $\frac{14}{22} - \frac{11}{22}$ **[1]** $= \frac{3}{22}$ **[1]**
   g) $\frac{27}{30} - \frac{20}{30}$ **[1]** $= \frac{7}{30}$ **[1]**
2. **a)** $\frac{4}{45}$ **[1]**
   b) $\frac{9}{70}$ **[1]**
   c) $\frac{10}{36} = \frac{5}{18}$ **[1]**
   d) $\frac{2}{9} \times \frac{4}{1}$ **[1]** $= \frac{8}{9}$ **[1]**
   e) $\frac{4}{5} \times \frac{11}{6} = \frac{44}{30}$ **[1]** $= \frac{22}{15} = 1\frac{7}{15}$ **[1]**
3. $\frac{2}{5} + \frac{1}{4} = \frac{8}{20} + \frac{5}{20}$ **[1]** $= \frac{13}{20}$
   $1 - \frac{13}{20}$ **[1]** $= \frac{7}{20}$ **[1]**
4. $\frac{4}{9} + \frac{1}{3} = \frac{7}{9}$ **[1]**
   $1 - \frac{7}{9} = \frac{2}{9}$ **[1]**
   Shared equally $= \frac{1}{9}$ chocolate **[1]**

1. **a)** $\frac{77}{9}$ **[1]**
   b) $\frac{23}{7}$ **[1]**
   c) $\frac{14}{11}$ **[1]**
2. **a)** $4\frac{3}{8}$ **[1]**
   b) $1\frac{11}{16}$ **[1]**

Page 75

1.

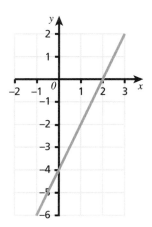

[2]

[1] for at least three correct values plotted.

2.

x	–3	–2	–1	0	1	2	3
y	–5	–5	–3	1	7	15	25

All correct **[3]**; At least five correct **[2]**; At least three correct **[1]**.

1. $y = x^2 - 4x + 6$
   $x = 3$: $3^2 - 4(3) + 6 = y = 3$. So, yes **[1]** (3, 3) is a coordinate on
   the graph as when $x = 3$, $y = 3^2 - 4(3) + 6 = 3$ **[1]**
2. $x = \frac{1}{2}$ **[1]**
   $y = 4$ **[1]**

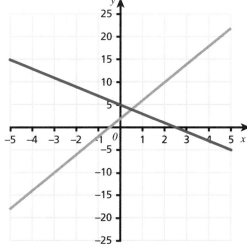

— $y = 4x + 2$ — $y = -2x + 5$ **[1]**
3. **A** and **E** **[1]**
   B and **D** **[1]**
   C and **F** **[1]**

Pages 76–77 **Practice Questions**

Page 76

1. $180 - 106 = 74$ **[1]**, $74 \times 2 = 148$, $180 - 148 = 32°$ **[1]**
2. **a)** $x = 134°$ **[1]** as alternate angle, $y = 180 - 134 = 46°$ **[1]**
   b) $x = 180 - 53 = 127°$ **[1]**, $y = 127°$ **[1]** as it is a corresponding
   angle
3. Decagon **[1]**
4. $180 - 150 = 30°$ (exterior angle) **[1]**
   $\frac{360}{30} = 12$ sides **[1]**

1.

x	–1	0	1	2	3
y	–6	–4	–2	0	2

[2]

[1] for at least three correct values.

1. a) $x = 180° - 130° = 50°$ **[1]**
   $y = 180° - 115° = 65°$ **[1]**
   b) $v = 140°$ **[1]**
   $w = 180° - 140° = 40°$ **[1]**

Page 77

1. $1 - 0.65$ **[1]** $= 0.35$ **[1]**

> Probability of all outcomes take away the probability of landing the other way up.

2. a)

Number	Frequency	Estimated probability
1	5	$\frac{5}{50} = \frac{1}{10}$
2	8	$\frac{8}{50} = \frac{4}{25}$
3	7	$\frac{7}{50}$
4	7	$\frac{7}{50}$
5	8	$\frac{8}{50} = \frac{4}{25}$
6	15	$\frac{15}{50} = \frac{3}{10}$
	Total 50	1

All correct **[3]**; At least five rows correct **[2]**; At least three rows correct **[1]**.

b) i) $\frac{3}{10}$ **[1]**

ii) $\frac{5}{50} + \frac{7}{50} + \frac{8}{50} = \frac{20}{50} = \frac{2}{5}$ **[1]**

iii) $\frac{15}{50} + \frac{8}{50} = \frac{23}{50}$ **[1]**

c) The dice is unlikely to be fair **[1]** as the probability of getting a 6 is a lot higher than the theoretical $\frac{1}{6}$ **[1]**

1. a)

Yellow	Red	Blue	Green	Pink
0.26	0.18	0.09	0.32	0.15

[1]

b) $1 - 0.26$ **[1]** $= 0.74$ **[1]**
c) $0.09 + 0.32 = 0.41$ **[1]**

Pages 78–85 Revise Questions

Page 79 Quick Test
1. 0.35 and 35%
2. $\frac{9}{25}$
3. £28
4. $49

Page 81 Quick Test
1. £405
2. £92 000
3. 84%
4. 80%
5. £460
6. £150 000

Page 83 Quick Test
1. ÷ 6
2. 12
3. 5
4. $y = 5$
5. $x = -9$

Page 85 Quick Test
1. $x = 5$
2. $x = 3$
3. $x = 2$
4. 11
5. 2 kg

Pages 86–87 Review Questions

Page 86

1. a) $x = 180 - 61 = 119°$
   $y = 119°$ (as it is corresponding) **[2]**
   b) $y = 180 - 114 = 66°$
   $x = 66°$ (as it is alternate angle) **[2]**
2. 1260° **[1]**
3. $180 - 160 = 20°$ **[1]**
   $\frac{360}{20}$ **[1]** = 18, so it is an 18-sided shape **[1]**

1. A regular pentagon has an interior angle of 108° **[1]**. 108 is not a factor of 360 **[1]** so therefore the tessellation would either create an overlap or a gap. **[1]**

2.

[2]

Page 87

1. a)

Sprinkles	Frequency	Probability
Chocolate	19	$\frac{19}{50} = 0.38$
Hundreds and thousands	14	$\frac{14}{50} = 0.28$
Strawberry	7	$\frac{7}{50} = 0.14$
Nuts	10	$\frac{10}{50} = 0.2$

All correct **[2]**; At least two rows correct **[1]**.

b) $\frac{19}{50} + \frac{10}{50}$ **[1]** $= \frac{29}{50}$ or 0.58 **[1]**

2. $1 - 0.47 = 0.53$ **[1]**

3. a)

Sales destination	Probability of going to destination
London	0.26
Cardiff	0.15
Chester	0.2
Manchester	0.39

[1]

b) Cardiff (it has the lowest probability) **[1]**

1. 0.68×325 **[1]** = 221 claims **[1]**
2. $1 - 0.14 = 0.86$ **[1]**
   $0.86 \times 250 = 215$ bread rolls that are good **[1]**

Page 88

1.

Fraction	Decimal	Percentage
$\frac{3}{5}$	0.6	60
$\frac{55}{100} = \frac{11}{20}$	0.55	55
$\frac{32}{100} = \frac{8}{25}$	0.32	32
$\frac{3}{100}$	0.03	3

[1]

[1]

[1]

[1]

2. **a)** $15 \div 3 \times 2$ [1]
    $= £10$ [1]
  b) $210 \div 7 \times 3$ [1]
    $= £90$ [1]
  c) $27 \div 9 \times 4$ [1]
    $= £12$ [1]

1. **a)** 10% of $80 \text{ cm} = 80 \div 10 = 8 \text{ cm}$
    $5\% = 8 \div 2 = 4 \text{ cm}$ [1]
    $15\% = 8 + 4 = 12 \text{ cm}$ [1]
  b) 10% of $160 \text{ m} = 160 \div 10 = 16 \text{ m}$
    $30\% = 16 \times 3 = 48 \text{ m}$ [1]
    $5\% = 16 \div 2 = 8 \text{ m}$ [1]
    $35\% = 48 + 8 = 56 \text{ m}$ [1]
  c) $10\% = 70 \div 10 = £7$ [1]
    $5\% = 7 \div 2 = £3.50$ [1]
2. 10% of $£75 = 75 \div 10 = £7.50$
  $20\% = £7.50 \times 2 = £15$ [1]
  Sale price $= £75 - £15$ [1]
    $= £60$ [1]
3. **a)** $300 \div 10 = 30$
    $5\% = £15$ [1]
    After two years $£15 \times 2 = £30$ [1]
    Total in account $= £330$ [1]
  b) $300 \div 10 = 30$
    $5\% = £15$ [1]
    After five years $£15 \times 5 = £75$ [1]
    Total in account $= £375$ [1]

Page 89

1. **a)** $2x - 5 = 3$
    $(+5)$ $2x = 8$ [1]
    $(\div 2)$ $x = 4$ [1]
  b) $3x + 1 = x + 7$
    $(-x)$ $2x + 1 = 7$
    (-1) $2x = 6$
    $(\div 2)$ $x = 3$ [1]
  c) $2(2x - 3) = x - 3$
       $4x - 6 = x - 3$
    $(-x)$ $3x - 6 = -3$
    $(+6)$ $3x = 3$
    $(\div 3)$ $x = 1$ [1]
  d) $\frac{3x + 5}{4} = 5$
    $(\times 4)$ $3x + 5 = 20$ [1]
    (-5) $3x = 15$
    $(\div 3)$ $x = 5$ [1]
2. $3n + 2 = 11$ [1]
  (-2) $3n = 9$ [1]
  $(\div 3)$ $n = 3$
3. $4n = 48$
  $n = 48 \div 4$
  $n = 12$, so 12 people at party [1]

1. $3(x + 1) = 2 + 4(2 - x)$
     $3x + 3 = 2 + 8 - 4x$
     $3x + 3 = 10 - 4x$ [1]
  $(+4x)$ $7x + 3 = 10$ [1]
  (-3) $7x = 7$
  $(\div 7)$ $x = 1$ [1]
2. $5(2a + 1) + 3(3a - 4) = 4(3a - 6)$
    $10a + 5 + 9a - 12 = 12a - 24$
       $19a - 7 = 12a - 24$ [1]
  $(-12a)$ $7a - 7 = -24$ [1]
    $(+7)$ $7a = -17$
    $(\div 7)$ $a = -2\frac{3}{7}$ [1]
3. $\frac{x}{5} = \frac{(x + 6)}{15}$ [1]
  $15x = 5(x + 6)$ or $3x = x + 6$ [1]
  $10x = 30$ or $2x = 6$ [1]
  $x = 3$ [1]
4. $\frac{14}{y} = \frac{7}{(y - 4)}$ [1]
  $14(y - 4) = 7y$ or $2(y - 4) = y$ [1]
  $14y - 56 = 7y$ or $7y = 56$ or $2y - 8 = y$ [1]
  $y = 8$ [1]
5. $4x + 10 = 6x + 6$ **[2]** for correct equation **OR [1]** for any correct
  perimeter.
  $(-4x)$ $10 = 2x + 6$ [1]
  (-6) $4 = 2x$
  $(\div 2)$ $x = 2$ [1]

Page 91 Quick Test
1. 6
2. **a)** **b)**  **c)**

3. Rectangle $3 \text{ cm} \times 6 \text{ cm}$

Page 93 Quick Test
1. Any three shapes exactly the same size
2. D
3. 4–5 m
4. A and B
5. Any two similar shapes

Page 95 Quick Test
1. **a)** 6 : 8 or equivalent   **b)** 8 : 6 or equivalent
2. **a)** 1 : 3 **b)** 7 : 1 **c)** 1 : 4

Page 97 Quick Test
1. 10 : 25 : 5
2. Sara £200, John £160
3. 28
4. £36
5. 4 days

Page 98

1. **a)** $\frac{13}{25} = \frac{52}{100} = 52\%$ [1]
  b) $0.375 = \frac{375}{1000}$ **[1]** $= \frac{3}{8}$ **[1]**
  c) $36\% = \frac{36}{100}$ **[1]** $= \frac{9}{25}$ **[1]**
2. **a)** $5 \div 5 \times 2$ [1]
    $= £2$ [1]
  b) $£5 - (£2.50 + £2)$ [1]
    $= £0.50$ or 50p [1]
3. **a)** 10% of $300 \text{ cm} = 300 \div 10 = 30 \text{ cm}$
    $20\% = 30 \times 2$ [1]
    $= 60 \text{ cm}$ [1]

b) 10% of £140 = 140 ÷ 10 = £14
   5% = 14 ÷ 2 = £7 [1]
   1% = 140 ÷ 100 = £1.40 [1]
   6% = £7 + £1.40 = £8.40 [1]
c) 10% of 2800 = 2800 ÷ 10 = 280 g [1]
   30% = 280 × 3 = 840 g [1]
   5% = 280 ÷ 2 = 140 g [1]
   35% = 840 + 140 = 980 g [1]

4. 10% of £90 = 90 ÷ 10 = £9
   5% = £9 ÷ 2 = £4.50 [1]
   15% = £9 + £4.50 = £13.50 [1]
   Sale price = £90 − £13.50
              = £76.50 [1]

5. a) 150 ÷ 100 × 6
      6% = £9 [1]
      Total in account after one year = £159 [1]
   b) 150 ÷ 100 × 6
      6% = £9 [1]
      After four years £9 × 4 = £36 [1]
      Total in account = £186 [1]

1. $\frac{3}{25}$ $\frac{7}{50}$ 16%   0.18   0.2
   All correct [2]; Three in correct order [1]

Page 99

1. a) $6x - 5 = 4x + 7$
      $(-4x)$ $2x - 5 = 7$
      $(+5)$ $2x = 12$ [1]
      $(÷2)$ $x = 6$ [1]
   b) $5(x + 2) = 2(x - 1)$
             $5x + 10 = 2x - 2$ [1]
      $(-2x)$ $3x + 10 = -2$ [1]
      (-10) $3x = -12$
      $(÷3)$ $x = -4$ [1]
   c) $3x - 1 = 4 - 2x$
      $(+2x)$ $5x - 1 = 4$ [1]
      $(+1)$ $5x = 5$ [1]
      $(÷5)$ $x = 1$ [1]
   d) $\frac{6x - 5}{4} = 7$
      $(×4)$ $6x - 5 = 28$ [1]
      $(+5)$ $6x = 33$
      $(÷6)$ $x = 5.5$ [1]

2. $56 - n = 29$ [1]
   $n = 56 - 29 = 27$
   27 chocolate bars were sold. [1]

3. $\frac{20n + 150}{4} = 50$ [1]
   $(×4)$ $20n + 150 = 200$ [1]
   (-150) $20n = 50$
   $(÷20)$ $n = 2.5$
   The builders worked for $2\frac{1}{2}$ hours. [1]

1. $4(x - 2) - 2(3 - 2x) = 5x + 1$
   $4x - 8 - 6 + 4x = 5x + 1$
   $8x - 14 = 5x + 1$ [1]
   $(-5x)$ $3x - 14 = 1$ [1]
   $(+14)$ $3x = 15$
   $(÷3)$ $x = 5$ [1]

2. $8 - 2a = 6a + 12$ [1]
   $(+2a)$ $8 = 8a + 12$
   (-12) $-4 = 8a$ [1]
   $(÷8)$ $-0.5 = a$ [1]

Page 100

1.

[2]
[1] for any correct reflection or at least two squares in the correct position.

1. a) See diagram below [2]
      [1] for any 90° rotation.
   b) See diagram below [2]
      [1] for any enlargement of shape A or B by a scale factor 3.

   c) A and B [1]
   d) A and C **OR** B and C [1]

2. 13 ÷ 2 [1] = 6.5 cm [1]

Page 101

1. 2 : 7 [1]
2. a) 1 : 3 [1]
   b) 40 minutes : 90 minutes [1]
      4 : 9 [1]
   c) 300p : 80p [1]
      15 : 4 [1]
3. 480 ÷ 12 = 40 [1]
   Ann: 4 × 40 = £160
   Ben: 5 × 40 = £200
   Cara: 3 × 40 = £120 [2]
4. 3 parts = £27
   1 part = 27 ÷ 3 = £9 [1]
   2 parts = £9 × 2 = £18 [1]
   Total sum of money = £27 + £18 = £45 [1]

1. Butter: 40 ÷ 6 × 15 = 100 g [1]
   Flour: 100 ÷ 6 × 15 = 250 g [1]
2. a) 1 man takes 8 × 10 = 80 days [1]
      10 men take 80 ÷ 10 = 8 days [1]
   b) 1 man takes 8 × 10 = 80 days [1]
      5 men take 80 ÷ 5 = 16 days [1]

Page 103 Quick Test
1. a) 40 km   b) 16 km
2. a) 19 miles   b) 25 miles
3. a) £3   b) £4
   c)

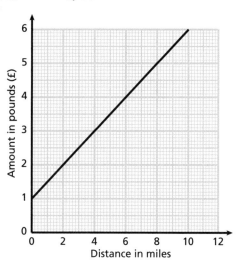

Page 105 Quick Test
1. 1.5 hours
2. 80 km/h
3. Kamala (21 km/h, John 20 km/h)
4. €240
5. 250 g

Page 107 Quick Test
1. a) 10.24   b) 244.9225
2. a) 70   b) 6.3
3. 5.46 cm
4. 7.94 cm

Page 109 Quick Test
1. a) 0.3420   b) 0.8660   c) 1
2. a) 56.5°   b) 55.2°   c) 88.2°

Page 110

1.

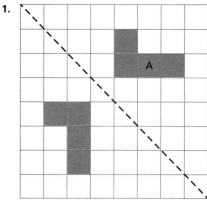

[2]

[1] for any correct reflection or at least two squares in the correct position.

---

1. a) See diagram [2]
   [1] for rotating shape A through 180° about any other point.
   b) See diagram [2]
   [1] for an enlargement in the wrong position.

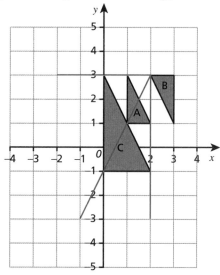

   c) Area shape A = 1 square unit [1]
   d) Area shape C = 4 square units [1]
   e) 1 : 4 [1]
2. 11.25 ÷ 9 = 1.25 [1]
   6.4 × 1.25 = 8 cm [1]

Page 111

1. a) 2 : 3 [1]
   b) 25p : 200p [1]
      1 : 8 [1]
   c) 750 m : 200 m [1]
      15 : 4 [1]
2. a) 40 : 15 [1]
   b) 7 : 12 [1]
   c) 6 : 7.5 [1]
3. 40 ÷ 8 = 5 [1]
   3 parts = 5 × 3 = 15
   5 parts = 5 × 5 = 25
   Ratio = 15 pens : 25 pens [1]
4. 4 parts = £120
   1 part = 120 ÷ 4 = £30 [1]
   3 parts = £30 × 3 = £90 [1]
   Altogether there is £30 + £120 + £90 = £240 [1]

---

1. a) In 1 hour the machine can make 1140 ÷ 8 = 142.5 cups [1]
   In 10 hours the machine can make 142.5 × 10 = 1425 cups [1]
   b) In 1 hour the machine can make 1140 ÷ 8 = 142.5 cups [1]
   In 12 hours the machine can make 142.5 × 12 = 1710 cups [1]
2. 1 man takes 3 × 2 = 6 days [1]
   3 men will take 6 ÷ 3 = 2 days [1]
3. 1 bag will feed 1 donkey for 2 days [1]
   4 bags will feed 4 donkeys for 2 days [1]
   10 bags will feed 4 donkeys for 5 days [1]
   OR
   In 1 day, 3 donkeys will eat 12 ÷ 8 = 1.5 bags of oats
   In 1 day, 1 donkey will eat 1.5 ÷ 3 = 0.5 bags of oats [1]
   In 1 day, 4 donkeys will eat 0.5 × 4 = 2 bags of oats [1]
   So 10 bags will last 4 donkeys 10 ÷ 2 = 5 days [1]

Page 112

1. Indicates France [1]
   and gives a correct justification
   Converts $ and € to £

   $1\,000\,000 \div 2.7 = £370\,370$ [1]
   $780\,000 \div 1.54 = £506\,494$
   OR
   Converts $ into €

   $1\,000\,000 \div 2.7 \times 1.54 = 570\,370$
   OR
   Converts € into $

   $780\,000 \div 1.54 \times 2.7 = 1\,367\,532$

2. Speed = distance ÷ time
   $= 350\,\text{km} \div 1.1\,\text{h}$ [1]
   $= 318\,\text{km/h}$ [1]

Be careful when using answers in further calculations.

1. a) Density = mass ÷ volume
      $= 2000 \div 0.5$ [1]
      $= 4000\,\text{kg/m}^3$ [1]

   b) Volume = mass ÷ density
      $= 5000\,\text{kg} \div 4000\,\text{kg/m}^3$ [1]
      $= 1.25\,\text{m}^3$ [1]

Page 113

1. a) $2^2 + 5^2 = BC^2$
      $BC = \sqrt{29}$ [1]
      $BC = 5.39\,\text{m}$ [1]

   b) $9^2 + AC^2 = 17^2$
      $AC = \sqrt{289 - 81} = \sqrt{208}$ [1]
      $AC = 14.4\,\text{cm}$ [1]

2. a) $\sin 25 = \frac{p}{17}$
      $p = \sin 25 \times 17$ [1]
      $p = 7.18\,\text{m}$ [1]

   b) $\tan y = \frac{32}{46}$
      $y = \tan^{-1}(32 \div 46)$ [1]
      $y = 34.8°$ [1]

---

1. $\cos x = \frac{8}{20}$ [1]
   $x = \cos^{-1}(8 \div 20)$ [1]
   $x = 66.4°$ [1]

2. a) No [1]
   b) Yes [1]
   c) No [1]
   d) Yes [1]

Page 114

1. a) $200 \div 1.75$ [1]
      $= £114.29$ [1]
   b) 200×1.75 [1]
      $= \text{US}\$350$ [1]

2. a) Time = $300 \div 40$ [1]
      $= 7.5$ hours [1]
   b) Distance = 40×4 [1]
      $= 160$ miles [1]

1. a)

Time (t minutes)	No. of bacteria
0	1
10	2
20	4
30	8
40	16
50	32
60	64
70	128
80	256
90	512
100	1024

[2]

[1] for 1, 2, 4, 8, 16, 32

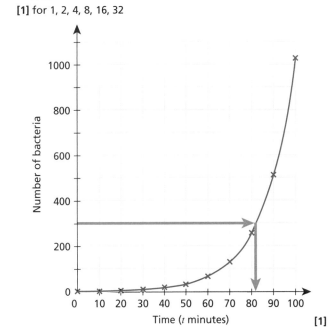

[1]

   b) 300 bacteria would take approximately 82 minutes [1]

Page 115

1. a) $\tan x = \frac{15}{8}$
      $x = \tan^{-1}(15 \div 8)$ [1]
      $x = 61.9°$ [1]
   b) $23^2 + 15^2 = x^2$
      $x = \sqrt{754}$ [1]
      $x = 27.5\,\text{m}$ [1]

2. a) $\sin 30° = \frac{3}{y}$
      $y = 3 \div \sin 30°$ [1]
      $y = 6\,\text{cm}$ [1]
   b) $6^2 + AB^2 = 9^2$
      $AB = \sqrt{81 - 36} = \sqrt{45}$ [1]
      $AB = 6.7\,\text{cm}$ [1]

3. $\sin 35° = \frac{\text{opp}}{18}$
   $\sin 35° \times 18 = \text{height}$ [1]
   Height of the pole = $10.3\,\text{m}$ [1]

1. a) $6.4^2 + h^2 = 15^2$
      $h = \sqrt{225 - 40.96} = \sqrt{184.04}$ [1]
      $h = 13.57\,\text{m}$ [1]
   b) $\cos x = \frac{6.4}{15}$
      $x = \cos^{-1}(6.4 \div 15)$ [1]
      $x = 64.7°$ [1]

Pages 116–121 No Calculator Allowed

1. a) Surface area = $2(6 \times 4 + 6 \times 2 + 4 \times 2) = 88\,cm^2$ **[1]**
   Volume = $6 \times 4 \times 2 = 48\,cm^3$ **[1]**
   b) Surface area = $2(12 \times 7 + 12 \times 8 + 7 \times 8) = 472\,cm^2$ **[1]**
   Volume = $12 \times 7 \times 8 = 672\,cm^3$ **[1]**

2. a) $4\frac{1}{2} + 2\frac{1}{3} = \frac{9}{2} + \frac{7}{3} = 6\frac{5}{6}$ **[1]**

   b) $5\frac{2}{3} + 8\frac{1}{4} = \frac{17}{3} + \frac{33}{4} = 13\frac{11}{12}$ **[1]**

   c) $9\frac{1}{6} - 2\frac{3}{8} = \frac{55}{6} - \frac{19}{8} = 6\frac{19}{24}$ **[1]**

   d) $12\frac{1}{2} - 14\frac{5}{6} = \frac{25}{2} - \frac{89}{6} = -2\frac{1}{3}$ **[1]**

3. a) **[1]**

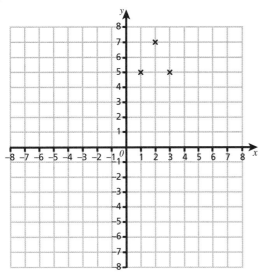

   b) (2, 3) **[1]**

4. a)

x	−2	−1	0	1	2	3
y	7	5	3	1	−1	−3

   All correct **[2]**; At least three correct **[1]**

   b)

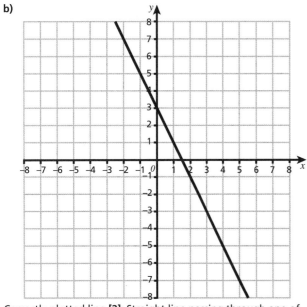

   Correctly plotted line **[2]**; Straight line passing through one of the correct coordinates **[1]**

5. a) $x = 115°$ (corresponding and opposite) **[1]**
   $y = 55°$ (alternate) **[1]**
   b) $x = 180 - 75 = 105°$ **[1]**
   $y = 180 - 85 = 95°$ **[1]**

6. a) i) $4x + 4y$ **[1]**
   ii) $3g + 1$ **[1]**
   b) i) $4x - 20$ **[1]**
   ii) $4x^2 + 16x$ **[1]**
   c) i) $6(x - 2)$ **[1]**
   ii) $4x(x - 2)$ **[1]**

7. $z = 2\,cm$ **[3]**
   [2] if 9 is seen; **[1]** for a correct attempt to find the area of the trapezium with no more than one numerical error

8. a) −2 and −3 **[1]**
   b) 8 and −1 **[1]**

9. a) $4c + 4b - 4a$ or equivalent **[2]**
   [1] if $b - a$ is seen
   b) 24 units **[2]**
   [1] for an attempt to substitute into answer from part **a)**

10. a) $400 + 40 = 440\,mm$ **[1]**
    b) $440 + 44 = 484\,mm$ **[1]**

11. a) $y = 5$ **[2]**
    [1] for $3y = 15$
    b) $x = 32$ **[2]**
    [1] for $12 - x = -20$ or $\frac{x}{4} = 8$

12.

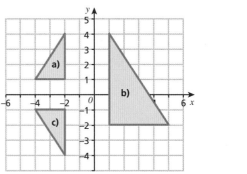

[3]

Pages 122–127 Calculator Allowed

1. a)

Type	Frequency	Probability	
Tug boat	12	$\frac{12}{40} = \frac{3}{10}$	**[1]**
Ferry boat	2	$\frac{2}{40} = \frac{1}{20}$	**[1]**
Sail boat	16	$\frac{16}{40} = \frac{2}{5}$	**[1]**
Speed boat	10	$\frac{10}{40} = \frac{1}{4}$	**[1]**

   b) $\frac{2}{5} \times 75 = 30$ **[1]**

2. a) Surface area = $326.7\,cm^2$ **[2]**
   [1] for $8 \times \pi \times 9 + 2(4^2 \times \pi)$
   Volume = $452.4\,cm^3$ **[2]**
   [1] for $4^2 \times \pi \times 9$
   b) Surface area = $298.5\,cm^2$ **[2]**
   [1] for $10 \times \pi \times 4.5 + 2(5^2 \times \pi)$
   Volume = $353.4\,cm^3$ **[2]**
   [1] for $5^2 \times \pi \times 4.5$

3. $43\,cm^2$ (to 2 d.p.) **[3]**
   [2] for 16 and $\pi(2^2)$ or 12.56637 seen;
   [1] for 16 or $\pi(2^2)$ or 12.56637 seen

4. Carol's Cars and £6000 and £6050 seen **[3]**
   [2] for £6000 and £6050 seen but no conclusion; **[1]** for £5400 seen

5. a) $\frac{(n+4)(n+5)}{2}$ **[1]**

   b) $\frac{(n+4)(n+5) + (n+3)(n+4)}{2} = n^2 + 8n + 16 = (n+4)^2$ **[3]**

   [2] for $n^2 + 8n + 16$ seen;

   [1] for $\frac{(n+4)(n+5) + (n+3)(n+4)}{2}$ seen

6. 10.6 m [2]
   [1] for sin 45° × 15

7.

Instruction	Mathematical expression
Think of a number	n
Multiply by 2	$2n$
Add 10	$2n + 10$
Divide by 2	$n + 5$
Subtract the number you thought of	5

All correct **[2]**; At least two correct **[1]**

8. $2x = 36 \times 3$ or $\frac{x}{3} = 36 \div 2$ or $x = \frac{36 \times 3}{2}$ **[1]**
   $x = 54$ **[1]**

9. a) 160 g of butter **[1]**, 400 g of flour **[1]**
   b) 200 g of butter **[1]**, 500 g of flour **[1]**

10. 19   [2]
   [1] for 96 or 77 seen

11. List numbers in order:
   10, 10, 14, 15, 20, 23, 25, 28, 28, 34, 34, 39, 45, 48, 49   **[1]**
   Median = 28 minutes   **[1]**

12. a) 7 inserted in the blank part of the circle for Set A   **[1]**
   b) 3   **[1]**
   c) $\frac{15}{30}$ or $\frac{1}{2}$ **[1]**

Glossary

a

additive identity   0 is called an additive identity as it leaves the answer of an addition or subtraction unchanged

alternate angles   angles that lie on opposite sides of a transversal and are equal in size

angle   the space (usually measured in degrees) formed by two intersecting lines or surfaces

arc   a curve; part of the circumference of a circle

area   the space inside a 2D shape

arithmetic sequence   a sequence of numbers with a common difference

associative law $(a + b) + c = a + (b + c)$

axis   a line that provides scale on a graph; often referred to as x-axis (horizontal) and y-axis (vertical)

b

bar chart   a chart that uses horizontal or vertical bars of equal width to represent statistics

biased   a statistical event where the outcomes are not equally likely

BIDMAS   the order in which operations should be carried out: Brackets; Indices; Division and Multiplication; Addition and Subtraction

binomial   an algebraic expression of the sum or difference of two terms

bisect   to cut exactly in two

bivariate data   data that compares the values of two variables

brackets   symbols used to enclose a sum

c

centre of enlargement   the position from which a shape is enlarged

centre of rotation   the point about which a shape is rotated

certain   an outcome of an event which must happen, probability equals 1

circle   a round 2D shape

circumference   the perimeter of a circle

class interval   in grouped data, the width of the group (difference between the upper and lower limit of the group)

coefficient   the number in front of a variable in a term, e.g. in the term $3a$ the coefficient is 3

combined events   when two or more events take place

common difference   the number added or subtracted at each stage of an arithmetic sequence

common factor   a number that can be divided into two different numbers, without leaving a remainder

commutative law $a + b = b + a$

composite   a complex 2D or 3D shape made from several simpler shapes

conditional probability   probability that is affected by an earlier outcome

congruent   exactly the same

constant   a value that does not change

conversion graph   a graph used to convert one unit to another

coordinates   usually given as (x, y); the x-value is the position horizontally, the y-value the position vertically

correlation   the relationship between data, the 'pattern'; can be positive or negative

corresponding angles   angles that are in the same position on two parallel lines relative to the transversal and are equal in size

cos (cosine)   the ratio of the adjacent side to the hypotenuse in a right-angled triangle

cross-section   a cross-section of a solid is a slice cut through the solid at right-angles to its axis

cube number   a number that can be expressed as the product of three equal integers

cube root   the inverse (or opposite) of cubing, i.e. the number that is multiplied by itself twice to form the cube

cylinder   a 3D shape with a circular top and base of the same size

d

data   a collection of answers or values linked to a question or subject

decimal   a number that contains tenths, hundredths, etc

decimal places   the number of digits after the decimal point

decrease   to make smaller

degree   a unit of measure of an angle

denominator   the bottom number of a fraction

density   the mass of something per unit of volume

diameter   the distance across a circle, going through the centre

difference   subtraction

direct proportion   quantities are in direct proportion if their ratio stays the same as the quantities increase or decrease

distance   length

distributive law $a(b + c) = ab + ac$

divide   to share

double   multiply by 2

e

edge   a line segment joining two vertices in a 2D or 3D shape

enlargement   a transformation in which a shape is made bigger or smaller

equally likely   having the same chance of happening

equation   a mathematical statement containing an equals sign

equivalent   the same as

estimate   a simplified calculation (not exact), often rounding to 1 significant figure

even chance   an equally likely chance of an event happening or not happening

event   a set of possible outcomes from a particular experiment

expand   remove brackets by multiplying

experimental probability   the ratio of the number of times an outcome happens to the total number of trials

exponent   another word for the 'power' or 'index'; see *index*

exponential graph   a graph of the form $y = k^x$

expression   a collection of algebraic terms

exterior angle   an angle outside a polygon formed between one side and the adjacent side extended

f

face   a side of a 3D shape

factor   a number that divides exactly into another number

factorise   take out the highest common factor and add brackets

fair   not biased

formula   a rule linking two or more variables

fraction   any part of a number or 'whole'

frequency table   a table that shows the number of times 'something' occurs

frequency diagram   also known as a frequency chart, this is a diagram that shows the frequency of a given event

function machine   a flow diagram that shows the order in which operations should be carried out

g

geometric sequence   a sequence of numbers in which each term is the product of the previous term multiplied by a constant value

gradient   the measure of steepness of a line

grouped data   data that has been collected in or sorted into groups

h

highest common factor   the highest factor two or more numbers have in common

hypotenuse   the longest side of a right-angled triangle

hypothesis   a prediction of an experiment or outcome

i

impossible   an outcome of an event which cannot happen, probability equals 0

improper fraction   a fraction where the numerator is larger than the denominator

income tax   tax paid on money earned

increase   to make bigger

index   the power to which a number is raised; in 2^4 the base is 2 and the index is 4

infinite   a word that describes a sequence that continues forever

integer   whole number

intercept   the point at which a graph crosses an axis

interest   a monetary amount added on to savings or loans

interior angle   the measure of an angle inside a shape

interpret   to describe the trends shown in a statistical diagram or statistical measure; the way in which a representation of information is used or surmised

inverse   the opposite of

inverse proportion   a relationship where one value increases as another value decreases so that their product is always equal

l

like terms   terms with the same variables

likely   a word used to describe a probability which is between evens and certain on a probability scale

line of best fit   the straight line (usually on a scatter graph) that represents the closest possible line to each point; shows the trend of the relationship

linear   in one direction, straight

locus   the only set of points that satisfies certain conditions

lowest common multiple   the lowest multiple two or more numbers have in common

lowest terms   a fraction or ratio in which the parts have no common factors

m

mean   a measure of average; sum of all the values divided by the number of values

median   a measure of average; the middle value when data is ordered

mixed number   a number with a whole part and a fraction

mode   a measure of average; the most common

multiplicative identity   1 is called a multiplicative identity as it leaves the answer to a multiplication or division unchanged

multiplier   a number that you are multiplying by

mutually exclusive   events or outcomes that cannot happen at the same time

n

negative   less than zero

net   a 2D representation of a 3D shape, i.e. a 3D shape 'unfolded'

nth term   see *position-to-term*

numerator   the top number of a fraction

o

ordinary number   a number not written in standard form

outcome   a possible result of a probability experiment

outlier   a statistical value which does not fit with the rest of the data

p

parallel   lines are parallel if they are always the same distance apart; this means they will never meet

parallelogram   a quadrilateral with two pairs of equal and opposite parallel sides and equal opposite angles

percentage   out of 100

perimeter   distance around the outside of a 2D shape

perpendicular   at 90° to

perpendicular bisector   the line which cuts another line in half and is at right angles to it

pi (π)   the ratio between the diameter of a circle and its circumference, approx. 3.142

pictogram   a frequency diagram in which a picture or symbol is used to represent a particular frequency

pie chart   a circular diagram divided into sectors to represent data, where the angle at the centre is proportional to the frequency

place value   indicates the value of the digit depending on its position in the number

position-to-term   a rule which describes how to find a term from its position in a sequence

positive   greater than zero

power   see *index*

prime   a number with exactly two factors, itself and 1

prime factor decomposition   the process of breaking a number down into a product of prime factors

prism   a 3D shape with uniform cross-section

probability   the likeliness of an outcome happening in a given event

probability scale   a scale to measure how likely something is to happen, running from 0 (impossible) to 1 (certain)

product   multiplication

protractor   a piece of equipment used to measure angles

Pythagoras' Theorem   in a right-angled triangle, the square on the hypotenuse is equal to the sum of the squares of the other two sides

q

quadratic   based on square numbers

quadratic equation   an equation where the highest power of x is x^2

quadrilateral   a four-sided 2D shape

quantity   an amount

r

radius   half the diameter; the measurement from the centre of a circle to the edge

random   each possible outcome is equally likely

range   the difference between the biggest and smallest number in a set of data

ratio a comparison of two amounts

raw data original data as collected

ray a line connecting corresponding vertices

reciprocal the inverse of any number except zero, e.g. the reciprocal of 2 is $\frac{1}{2}$ and the reciprocal of $\frac{3}{4}$ is $\frac{4}{3}$

reciprocal graph a graph of the form $y = \frac{1}{x}$

rectilinear a word that describes shapes with straight edges and right angles

reflection a mirror image

regular polygon a 2D shape that has equal-length sides and equal angles

rotation a turn

rounding a number can be rounded (approximated) by writing it to a given number of decimal places or significant figures

s

sample space a way in which the outcomes of an event are shown

scale the ratio between two or more quantities

scale factor the ratio by which a shape/number has been increased or decreased

scatter graph paired observations plotted on a 2D graph

sector a section of a circle enclosed between an arc and two radii (a pie piece)

sequence a set of numbers or shapes which follow a given rule or pattern

share to divide

significant figures the importance of digits in a number relative to their position; in 3456 the two most significant figures are 3 and 4

similar two shapes that have the same shape but not the same size

simplify make simpler, normally by cancelling a fraction or ratio or by collecting like terms

simultaneous equations equations that represent lines that intersect

sin (sine) the ratio of the opposite side to the hypotenuse in a right-angled triangle

solve work out the value of

speed how fast something is moving

square a regular four-sided polygon; to multiply by itself

square number a number made from multiplying an integer by itself

square root the opposite of squaring; a number when multiplied by itself gives the original number

standard form a way of writing a large or small number using powers of 10, e.g. 120 000 = 1.2×10^5

subject a single variable isolated on one side of a formula

substitute to replace a letter in an expression with a number

sum addition or total

supplementary angles two angles that add up to 180°

surface area the total area of all the faces of a 3D shape

survey a set of questions used to collect information or data

t

tally chart a simple way of recording and showing information by using vertical marks to count each occurrence; every fifth tally mark is a diagonal line through the previous four

tan (tangent) the ratio of the opposite side to the adjacent side in a right-angled triangle

term a number that forms part of a sequence; in expressions, terms are separated by + and − signs

terminating decimal a decimal number with a finite number of digits after the decimal point, e.g. 0.75, 0.125

term-to-term the rule which describes how to move between consecutive terms

tessellation a pattern made by repeating 2D shapes with no overlap or gap

transversal a line that crosses two parallel lines

trapezium a quadrilateral with only one pair of parallel sides

triangle a three-sided 2D shape

triangular numbers numbers that can be represented by a triangular pattern of dots, e.g. 10

u

units these define length, speed, time, volume, etc

unknown a number that is not known

unlikely a word used to describe a probability which is between evens and impossible on a probability scale

v

value added tax (VAT) tax paid when certain items are purchased

variable a quantity that can take on a range of values, usually denoted by a letter such as x or y

vertex the point where two or more edges meet on a 2D or 3D shape

vertical line graph a diagram that is similar to a bar chart but uses lines rather than bars

vertically opposite angles equal angles that face each other at a point

volume the capacity, or space, inside a 3D shape

Index

Collins

KS3
Maths
Higher Level
Workbook

Trevor Senior

Contents

Rethink Revision

Have you ever taken part in a quiz and thought '*I know this*!', but no matter how hard you scrabbled around in your brain you just couldn't come up with the answer?

It's very frustrating when this happens, but in a fun situation it doesn't really matter. However, in tests and assessments, it is essential that you can recall the relevant information when you need to.

Most students think that revision is about making sure you **know** *stuff*, but it is also about being confident that you can **retain** that *stuff* over time and **recall** it when needed.

Revision that Really Works

Experts have found that there are two techniques that help with *all* of these things and consistently produce better results in tests and exams compared to other revision techniques.

Applying these techniques to your KS3 revision will ensure you get better results in tests and assessments and will have all the relevant knowledge at your fingertips when you start studying for your GCSEs.

It really isn't rocket science either – you simply need to:
- **test yourself** on each topic as many times as possible
- **leave a gap** between the test sessions.

It is most effective if you leave a good period of time between the test sessions, e.g. between a week and a month. The idea is that just as you start to forget the information, you force yourself to recall it again, keeping it fresh in your mind.

Three Essential Revision Tips

1 **Use Your Time Wisely**
- Allow yourself plenty of time
- Try to start revising six months before tests and assessments – it's more effective and less stressful
- Your revision time is precious so use it wisely – using the techniques described on this page will ensure you revise effectively and efficiently and get the best results
- Don't waste time re-reading the same information over and over again – it's time-consuming and not effective!

2 **Make a Plan**
- Identify all the topics you need to revise (this Complete Revision & Practice book will help you)
- Plan at least five sessions for each topic
- A one-hour session should be ample to test yourself on the key ideas for a topic
- Spread out the practice sessions for each topic – the optimum time to leave between each session is about one month but, if this isn't possible, just make the gaps as big as realistically possible.

3 **Test Yourself**
- Methods for testing yourself include: quizzes, practice questions, flashcards, past papers, explaining a topic to someone else, etc.
- This Complete Revision & Practice book gives you seven practice test opportunities per topic
- Don't worry if you get an answer wrong – provided you check what the right answer is, you are more likely to get the same or similar questions right in future!

Visit our website to download your free flashcards, for more information about the benefits of these revision techniques and for further guidance on how to plan ahead and make them work for you.

collins.co.uk/collinsks3revision

Number

MR **1** A two-digit cube number and a square number have a difference of 2

What are the two numbers?

.... and .... [2]

PS **2** Write down two numbers that are factors of 24 that are also factors of 27

.... and .... [2]

3 **a)** Place ticks in the cells of the table to show if the number is a multiple of 2, 3, 4 or 5

	Multiple of 2	Multiple of 3	Multiple of 4	Multiple of 5
80				
81				
82				

[4]

b) Find a number that is a common multiple of 2, 3, 4 and 5 .... [2]

MR **4** Decide whether each statement is **always true**, **sometimes true** or **never true**.

a) Multiples of 3 are also multiples of 9 .... [1]

b) Multiples of 12 are also multiples of 6 .... [1]

c) Adding three consecutive even numbers
will give a multiple of 4 .... [1]

PS **5** A red car and a blue car go around a toy track. They both start from the same point.
The red car takes 8 seconds to complete a circuit. The blue car takes 10 seconds.

After how many seconds will they next pass the start point at the same time?

.... [2]

6 The Venn diagram shows the prime
factors of 84 and 108

What is the highest common factor of 84 and 108?
Show how you worked out your answer.

Prime factors of 84 Prime factors of 108

2^2 7 3^2 3

.... [2]

Total Marks .... / 17

7 Work out each of the following.

a) $(12 + 8) \times (12 - 8)$

_____ [2]

b) $4 + 7^2 - 3^3$

_____ [2]

c) $(17 - 7)^2 - (18 - 10)^2$

_____ [2]

(PS) **8** The number 144 can be written in the form $2^x \times 3^y$

Work out the values of x and y.

$x =$ _____ , $y =$ _____ [2]

(MR) **9** Decide whether each statement is **true** or **false**. Give a reason for each answer.

a) A square number **cannot** be a prime number.

_____ [1]

b) A prime number **cannot** be a cube number.

_____ [1]

c) A square number **cannot** be a cube number.

_____ [1]

Total Marks _____ / 11

Sequences

(MR) **1**   Here are four sequences.

    A   10, 20, 30, 40, …   **B**   100, 95, 90, 85, …

    C   0.1, 0.01, 0.001, 0.0001, …   **D** $\frac{1}{9}, \frac{1}{7}, \frac{1}{5}, \frac{1}{3}, \dots$

    a)   Which of the sequences will contain the number 80?

                                                                                    [1]

    b)   Which of the sequences will contain negative numbers?

                                                                                    [1]

    c)   What position will 50 be in sequence B?

                                                                                    [2]

(MR) **2**   Here are the nth terms of eight sequences.

    Which of the sequences decrease?

    A $3n + 2$ **B** $3n - 2$ **C** $-3n + 2$ **D** $-3n - 2$

    E $\frac{1}{3}n + 2$ **F** $\frac{1}{3}n - 2$ **G** $-\frac{1}{3}n + 2$ **H** $-\frac{1}{3}n - 2$

                                                                                    [2]

3   Here is a sequence.   5, 9, 13, 17, 21, …

    Jack thinks the 10th term of this sequence is 42. Explain why he is **not** correct.

                                                                                    [2]

4   Work out the nth term of the sequence   3, 9, 15, 21, …

                                                                                    [2]

Total Marks   **/ 10**

MR **5** Here is a sequence of numbers.

0, 3, 8, 15, 24, 35, …

a) Work out the nth term.

.... [2]

b) Which of the numbers 48, 63 and 82 are in the sequence?
Explain how you know.

.... [2]

6 Write down the first four terms of the sequence with nth term $3n^2 + 1$

.... [2]

PS **7** Here are some terms of the Fibonacci sequence. 1, 1, 2, 3, 5, 8, 13, 21, …

You can use two consecutive terms of the Fibonacci sequence to obtain an approximate conversion from miles to kilometres. For example, 5 miles ≈ 8 km, 8 miles ≈ 13 km

a) Use this method to convert 21 miles to kilometres.

.... km [1]

b) Convert 15 miles to kilometres.

.... km [1]

c) Convert 63 kilometres to miles.

.... miles [1]

8 Write down the nth term of the sequence 1, 8, 27, 64, …

.... [1]

Total Marks .... / 10

Perimeter and Area

1 Here are five triangles.

Which of the triangles have the same area as triangle A?

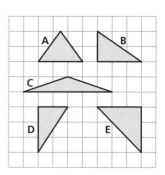

_____ [2]

2 Two triangles X and Y are joined together to form a trapezium as shown.

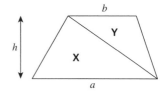

a) Write down a formula for the area of triangle X.

_____ [1]

b) Write down a formula for the area of triangle Y. _____ [1]

c) Use your answers to parts a) and b) to show that the area of the trapezium is $\frac{1}{2}(a + b) \times h$

_____

_____ [1]

3 Use the fact that a circle with radius 5 cm has circumference 31.4 cm.

a) What is the circumference of a circle of radius 15 cm?

_____ cm   [2]

b) What is the perimeter of a semicircle of radius 5 cm?

_____ cm   [2]

4 a) Work out the area of a circle of diameter 10 cm.

_____ cm²   [2]

b) Work out the unshaded area.

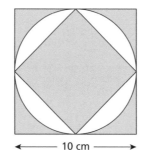

10 cm

_____ cm²   [3]

Total Marks _____ / 14

5 The outer diameter and the inner diameter of this sign are shown.

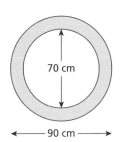

Work out the shaded area.
Give your answer to the nearest 100 cm²

_____ cm²   [4]

(PS) 6 This garden has three sections. The shapes of each section are: an isosceles trapezium, a right-angled triangle and a quarter circle.

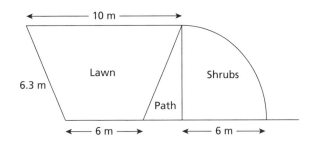

a) Work out the area of each section.

Lawn: _____ m²

Shrubs: _____ m²

Path: _____ m²   [6]

b) Work out the perimeter of the garden.

_____ m   [3]

Total Marks _____ / 13

Statistics and Data

1 Each week for four weeks, Anji spends £40 on fuel. On the fifth week, she spends £60.

What is the mean amount she spent on fuel over the five weeks?

£ .... [3]

(MR) **2** The table shows the number of tickets sold for a pantomime.

	Stall	Circle	Balcony	Total
Adult	240		192	
Child	160	147		
Total		300		1000

a) Complete the table. [2]

b) How many more adult tickets were sold than child tickets?

.... [2]

c) Adult tickets cost £25 and child tickets cost £20

How much money was taken from selling tickets?

£ .... [2]

3 The table shows information about the number of text messages sent by 30 students in one day.

Number of texts	Frequency
0–9	16
10–19	6
20–29	7
30 or more	1

a) What is the modal class? .... [1]

b) One of those students sent over 200 texts.

Why is the mean **not** suitable to use as an average for this data?

.... [1]

Total Marks .... / 11

KS3 Maths Workbook

4 The vertical line graph shows the heights of five plants, A, B, C, D and E.

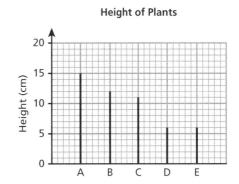

Height of Plants

a) Work out the mean height of the plants.

.... cm   [2]

b) An extra plant is now included. The mean height of all six plants is 11 cm.

What is the height of the extra plant?   .... cm   [2]

PS **5** The median of these five numbers is 7

| $x+2$ | | $x+4$ | | $x-3$ | | $x-1$ | | $x+4$ |

a) Work out the value of x.

$x =$ ....   [2]

b) Work out the value of the mode.   ....   [1]

MR **6** The table shows the times for 50 cyclists to complete a time trial.

Time, t (seconds)	Frequency, f
$10 < t \leqslant 12$	8
$12 < t \leqslant 14$	25
$14 < t \leqslant 16$	14
$16 < t \leqslant 18$	3

a) Work out the greatest possible range of the times.

.... s   [1]

b) The actual range of the times was 4.9 seconds. The fastest time was 11.6 seconds.

What was the slowest time?

.... s   [2]

Total Marks .... / 10

Decimals

(MR) **1** Show that this statement is correct. $5 \times (13 \times 2.6 - 13 \times 0.6) = 10^2 + 10 \times 3$

....

....

[3]

2 **a)** Round 83 to the nearest 10 .... [1]

b) Round 83 to 1 significant figure. .... [1]

c) Round 274 to the nearest 100 .... [1]

d) Round 274 to 1 significant figure. .... [1]

e) I am thinking of a number. The number rounded to 1 significant figure is 6000

What is the smallest number I could be thinking of? .... [1]

3 Complete the table by rounding each number to the given number of significant figures.

	1 s.f.	2 s.f.	3 s.f.	4 s.f.
1407				
2999				

[4]

4 Estimate the answer to each of the following.

a) 7.93×0.21

.... [2]

b) $\dfrac{30.1 + 8.9}{2.95}$

.... [2]

c) $\dfrac{19.1}{0.49}$

.... [2]

d) $(40.1)^2$

.... [2]

5 Estimate the perimeter and the area of a square with side 38.9 cm.

Perimeter: .... cm

Area: .... cm^2 [4]

6 Given that $6.4 \times 1.5 = 9.6$, work out:

a) 12.8×1.5

.... [2]

b) 6.4×2.5

.... [2]

Total Marks .... / 28

7 Write each of these numbers as an ordinary number.

a) 4.7×10^3 .... [1]   b) 8.31×10^2 .... [1]

c) 6.0×10^{-1} .... [1]   d) 1.35×10^{-4} .... [1]

8 The population of Australia is 26 002 413. The total land area is 7.7×10^6 km^2

Work out the average number of people per square kilometre in Australia.
Give your answer to the nearest whole number.

.... people/km^2 [2]

9 Convert the numbers into standard form and write them from smallest to biggest.

5678   0.000 456   7.89   12 345   0.234

.... [3]

10 The answer to a calculation is given as 3.7 (to 1 decimal place).

Complete the error interval.   .... ⩽ error < .... [2]

Total Marks .... / 11

Algebra

1   d is the number of days and h is the number of hours.

Which of the following shows the relationship between the number of hours and the number of days? Explain your answer.

$d = 24h$ 　　　　　 $d = 7h$ 　　　　　 $h = 7d$ 　　　　　 $h = 24d$

_____

_____ [2]

2   a) Complete this algebraic multiplication table.

×	3a	4
5a		
4	12a	

[2]

b) Use your answer to part a) to multiply out and simplify $5a(3a + 4) + 4(3a + 4)$

_____ [1]

3   Use the distributive law to write equivalent expressions for each of the following.

a) $3(x + 8)$ 　_____ [1] 　　b) $30(x + 8)$ 　_____ [1]

c) $3a(x + 8)$ 　_____ [1] 　　d) $3ab(x + 8)$ 　_____ [1]

4   Multiply out and simplify:

a) $(a + 3)(a + 4)$ _____ [2]

b) $(b - 2)(b + 6)$ _____ [2]

c) $(c + 5)^2$ _____ [2]

MR 5   Andy wants to work out the product $(3x + 4)(2x + 5y + 6)$

This is his answer.

What has Andy done wrong? Complete his answer.

×	3x	4
2x	$6x^2$	8x
5y	15xy	20y
6	18x	24

_____

_____ [1]

Total Marks _____ / 16

6 Expand and simplify $(4x + 1)(2x - 5)$

[2]

7 Expand and simplify $(x + 1)(x + 2)(x + 3)$

[3]

8 This formula is used in financial mathematics. $I = \dfrac{PRT}{100}$

Complete these rearrangements of the formula.

a) $PRT =$ .... [1]

b) $PR =$ .... [1]

c) $P =$ .... [1]

9 This formula is used in science. $s = \dfrac{1}{2}(u + v)\,t$

a) Work out the value of s when $u = 3$, $v = 7$ and $t = 5$

[2]

b) Work out the value of v when $s = 9$, $u = 4$ and $t = 6$

[3]

Total Marks .... / 13

3D Shapes: Volume and Surface Area

 1

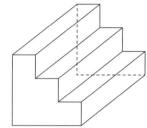

a) How many of the faces of the prism are rectangular? _____ [1]

b) How many faces does the prism have in total? _____ [1]

2 a) Draw labelled sketches of all the faces of this prism.

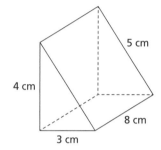

5 cm

4 cm

8 cm

3 cm

[4]

b) Use your sketches to work out the total surface area of the prism.

_____ cm² [4]

PS **3** A cylinder and a cuboid are shown.

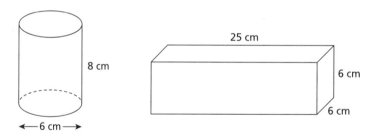

25 cm

8 cm

6 cm

6 cm

6 cm

a) Work out the volume of the cylinder.
Give your answer to 1 decimal place.

_____ cm³ [2]

b) Work out the total surface area of the cylinder.
Give your answer to 1 decimal place.

_____ cm² [2]

c) How many of the cylinders will fit inside the cuboid?

_____ [1]

Total Marks _____ / 15

4 **a)** Work out the volume of the triangular prism.

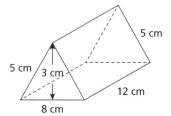

_____ cm³   [3]

b) Work out the total surface area of the triangular prism.

_____ cm²   [2]

MR **5** Euler's formula states that $F + V = E + 2$ where F is the number of faces, V is the number of vertices and E is the number of edges.

Show that this is true for:

a) a cuboid

_____

_____   [2]

b) a triangular prism

_____

_____   [2]

c) a square-based pyramid

_____

_____   [2]

PS **6** Work out the volume of the T-shaped prism.

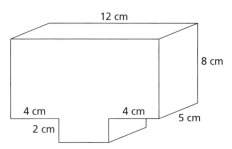

_____ cm³   [3]

Total Marks _____ / 14

Interpreting Data

MR **1** The bar chart shows the number of students in four year groups in a school.

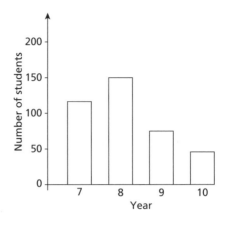

a) Which year group is the mode? .... [1]

b) If Year 11 was included, Year 11 would be the mode.

What can you say about the number of students in Year 11?

.... [1]

2 The pie chart shows the number of pieces of homework that 24 students are given one day.

Work out the mean number of pieces of homework per student.

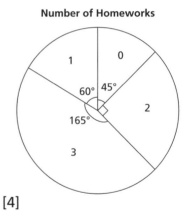

Number of Homeworks

.... [4]

MR **3** The table shows the age of some randomly selected animals on two farms.

Farm A	8	7	8	9	7	6	5	9	9	7	8	5
Farm B	4	10	8	8	8	9	11	10	8	8	10	8

Ted says that farm B has a bigger range so the ages are less consistent. Is he correct?

Show your working, commenting on any outliers.

....

.... [2]

Total Marks .... / 8

4 Here are some test results for 10 students.

Student	A	B	C	D	E	F	G	H	I	J
Mathematics	28	32	50	45	36	42	18	24	21	47
Science	25	27	50	35	37	35	15	19	22	40

a) Plot the test scores on the grid and draw a line of best fit. [3]

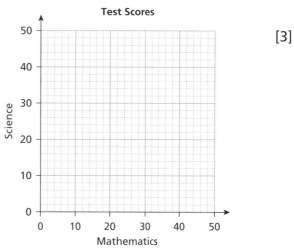

b) One student only did the maths test. Her score was 34.

Use your line of best fit to estimate her science score. .... [1]

(MR) **5** The graphs show the values of two cars over seven years.

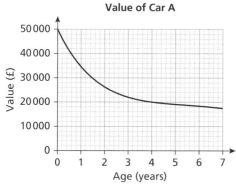

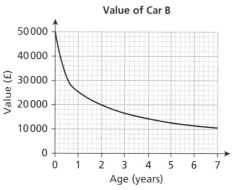

a) Which car had the greater fall in value? Give a reason for your answer.

.... [1]

b) Which car lost 30% of its value in the first year? .... [1]

c) What fraction of its original value
was car B worth after seven years? .... [1]

Total Marks .... / 7

Fractions

(MR) **1** The diagram shows how $\frac{1}{2} \times \frac{1}{5} = \frac{1}{10}$

Use this grid to work out $\frac{1}{2} \times \frac{3}{5}$

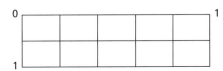

.... [2]

2 Insert $<$, $>$ or $=$ into each box to make the statements correct.

a) $5 \times \frac{1}{3}$ ☐ $\frac{5}{3}$ [1]

b) $\frac{1}{3} \times 4$ ☐ $1\frac{2}{3}$ [1]

c) $\frac{1}{5}$ ☐ $\frac{1}{2} \times \frac{1}{3}$ [1]

d) $\frac{1}{3} \times \frac{1}{2}$ ☐ $\frac{2}{5}$ [1]

3 Work out:

a) $\frac{1}{3} \times \frac{1}{4}$ .... [1]

b) $5 \times \frac{1}{3} \times \frac{1}{4}$ .... [1]

c) $\frac{2}{3} \times \frac{3}{4}$ .... [1]

d) $\frac{2}{3} \times \frac{3}{4} \times \frac{4}{5}$ .... [1]

(MR) **4** Use the fact that $\frac{7}{9} \times 234 = 182$ to work out:

a) $\frac{14}{9} \times 234$ .... [2]

b) $\frac{7}{3} \times 234$ .... [2]

c) $\frac{7}{18} \times 234$ .... [2]

5 Matt is flying to Malaga in Spain. The flight will take $2\frac{1}{2}$ hours. He checks the time and sees he is 45 minutes into the flight.

How long is it to the end of the flight?
Give your answer as a mixed number in hours.   .... hours [2]

Total Marks .... / 18

PS **6** Answer the following.

a) Work out $1\frac{1}{3} + 2\frac{3}{4} + 1\frac{1}{2}$

.... [3]

b) Use the fact that $4\frac{5}{8} + 6\frac{2}{3} = 11\frac{7}{24}$ to write down the answer to $1\frac{5}{8} + 7\frac{2}{3}$

.... [1]

c) Use the fact that $4\frac{5}{8} + 6\frac{2}{3} = 11\frac{7}{24}$ to write down the answer to $11\frac{7}{24} - 8\frac{2}{3}$

.... [1]

MR **7** Here is a sequence of numbers.   2   3 $4\frac{1}{2}$

a) Complete the sentence.

To find the next number you multiply the previous number by ....   [1]

b) Work out the next number in the sequence.

.... [2]

MR **8** Tim and Aisha are working out $3\frac{1}{2} \times 2\frac{1}{2}$

Tim says, "I did $3 \times 2 = 6$ and $\frac{1}{2} \times \frac{1}{2} = \frac{1}{4}$, so my answer is $6\frac{1}{4}$"

Aisha says, "$3\frac{1}{2} \times 2 = 7$ and $3\frac{1}{2} \times \frac{1}{2} = 1\frac{3}{4}$, so Tim must be wrong as the answer must be greater than 7."

a) Who is correct?

.... [1]

b) What is the correct answer?

.... [2]

Total Marks .... / 11

Coordinates and Graphs

(PS) 1   Here are the coordinates of three points.   (4, 11)   (5, 13)   (7, 17)

     a)   Work out the relationship between the x and y values for this set.

       .... [2]

     b)   Does a straight line pass through all three points?

       .... [1]

(MR) 2   Here are some coordinates.   (−5, −5) (−1, 3) (4, 13)

     Circle the correct equation of the line passing through all three coordinates.

     A $y = x$ **B** $y = 3x + 6$ **C** $y = 3x + 1$ **D** $y = 2x + 5$ [1]

(MR) 3   The points (−4, 2), (−2, 4) and (8, 14) lie on a straight line.

     Ethan thinks the equation of the line passing through the coordinates is $x − 6 = y$
     Explain why Ethan is wrong.

     .... [2]

4   **a)**   Complete the table for $y = x^2 - 2$

x	−3	−2	−1	0	1	2	3
y	7				−1	2	

       [2]

     b)   On the grid, draw the graph of
       $y = x^2 - 2$ from $x = -3$ to $x = 3$

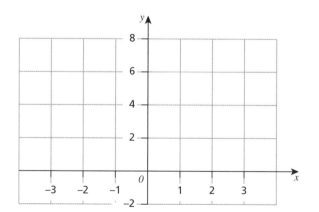

       [2]

Total Marks .... / 10

5 **a)** Draw the graphs of these simultaneous equations on the grid.

$y = x + 3$ and $y = 5 - x$

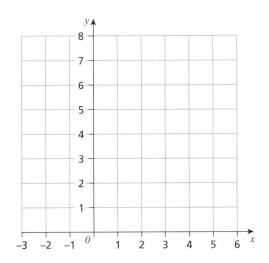

[2]

b) Use the grid to find the solution to the simultaneous equations.

$x = $ .... , $y = $ .... [1]

6 The graph of $y = 2x - 1$ is shown on the grid for values of x from -1 to 2

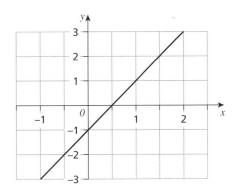

a) Use the graph to solve the equation $2x - 1 = 0$

$x = $ .... [1]

b) Use the graph to solve the equation $2x - 1 = 2$

$x = $ .... [1]

Total Marks .... / 5

Angles

1 $a = 35°$. Work out the size of angles b and c. Give a reason for your answers.

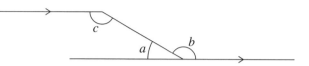

$b =$ _____ °   Reason: _____

$c =$ _____ °   Reason: _____ [2]

2   Work out the size of all the lettered angles.

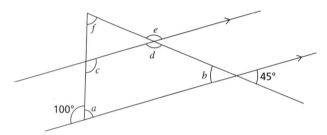

$a =$ _____ ° $b =$ _____ ° $c =$ _____ °

$d =$ _____ ° $e =$ _____ ° $f =$ _____ ° [6]

(MR) **3**   Circles intersect as shown. A quadrilateral is formed using the centres and the points of intersection as vertices. In each case give the special name of the quadrilateral.

a) These circles have the same radii.   **b)**   **c)**

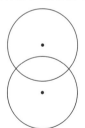
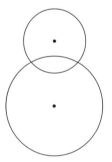
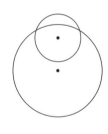

_____   _____   _____ [3]

4   The diagram shows a regular octagon.

By splitting the octagon into triangles, work out the sum of the interior angles.

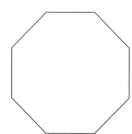

_____ ° [2]

Total Marks _____ / 13

(MR) 5 The diagram shows a regular pentagon.

Work out the sizes of angles a, b and c.
Give reasons for your answers.

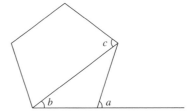

$a =$ .... ° Reason ....

$b =$ .... ° Reason ....

$c =$ .... ° Reason .... [6]

6 Work out the size of angle x in this hexagon.

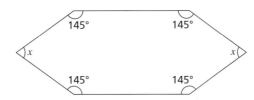

$x =$ .... ° [2]

7 The interior angle of a regular polygon is 135°

How many sides does the polygon have?

.... [2]

(PS) 8 The diagram shows a parallelogram inside a rectangle.

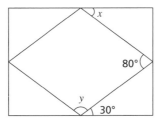

Work out the size of angles x and y.

$x =$ .... °

$y =$ .... ° [2]

Total Marks .... / 12

Probability

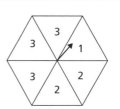

1. A fair six-sided spinner is shown. The arrow is spun.

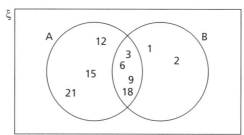

a) What is the probability that the arrow lands on 1? .... [1]

b) What is the probability that the arrow lands on 2? .... [1]

c) What is the probability that the arrow lands on 3? .... [1]

d) The spinner is now spun 50 times.

How many times do you expect the spinner to land on 3? .... [1]

2. Here is a Venn diagram.

a) Write down the numbers that are in set A. .... [1]

b) Write down the numbers that are in set B'. .... [1]

c) Write down the numbers that are in set A ∪ B. .... [1]

d) Describe the numbers that are in set B.

.... [1]

e) A number is chosen at random from the Venn diagram.

What is the probability that it is in set A ∩ B? .... [1]

3. Two fair coins are thrown. They land on either heads (H) or tails (T).

a) List all the possible outcomes. The first one is done for you.

HH .... [1]

b) What is the probability of two heads? .... [1]

Total Marks .... / 11

PS **4** A bag of money contains three 50p coins, fifteen £1 coins and twelve £2 coins.
A coin is chosen at random.

a) What is the probability that it is a £2 coin?
Give your answer as a fraction in its simplest form.

.... [2]

b) What is the probability that the coin is worth less than £2?
Give your answer as a fraction in its simplest form.

.... [2]

5 A box contains 300 toy bricks of different colours. 80 of the bricks are red, 100 are blue,
and the rest are yellow or green. There are twice as many yellow bricks as green bricks.

A brick is chosen at random.

a) What is the probability that the brick is blue?
Give your answer as a fraction in its simplest form.

.... [2]

b) What is the probability that the brick is **not** red?
Give your answer as a fraction in its simplest form.

.... [2]

c) What is the probability that the brick is yellow?
Give your answer as a fraction in its simplest form.

.... [3]

d) Ten bricks of each colour are now removed.

What is the probability that a brick chosen at random is now blue?

.... [2]

Total Marks .... / 13

FS **1** **a)** Increase £40 by 30%   £ .... [2]

b) Increase £30 by 40%   £ .... [2]

c) Decrease £50 by 20%   £ .... [2]

d) Decrease £60 by 70%   £ .... [2]

FS **2** **a)** Match the **four** statements to the correct calculation.   [4]

Increase £40 by 15%	Increase £15 by 40%	Decrease £40 by 15%	Decrease £15 by 40%

£15 × 1.4	£40 × 1.5	£40 × 1.15	£15 × 0.6	£40 × 0.6	£40 × 0.85

b) For the two calculations that have not been used, write a matching statement.

....

.... [2]

3 What is 15 out of 25 as a percentage?

.... % [2]

FS **4** A savings account pays 4% simple interest per year.

a) Jon pays £500 into the account.

How much interest will he be paid after 1 year?

£ .... [2]

b) How much will be in the account after 3 years?

£ .... [2]

Total Marks .... / 20

5 The cost of a dress is reduced from £75 to £30

Work out the percentage reduction.

.... %   [2]

(FS) **6** Anuja earns £45 000 per year. The first £12 500 is tax free. She pays 20% income tax on the rest.

a) How much income tax does she pay?

£ .... [3]

b) Her pay after tax is paid in 12 monthly instalments.

How much is each instalment? Give your answer to the nearest £100

£ .... [3]

7 Electricity costs 28p per kWh. There is also a standing charge of 48p per day.
VAT is added to the total at 5%

A household uses 250 kWh of electricity over 30 days.

Complete the table to work out the total bill.

Energy used	250 kWh @ 28p per kWh	£ ....
Standing charge	30 days @ 48p per day	£ ....
Subtotal of charges before VAT		£ ....
VAT @ 5%		£ ....
Total electricity charges		£ .... [5]

Total Marks .... / 13

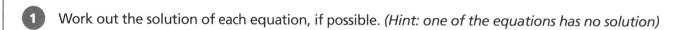
Equations

1 Work out the solution of each equation, if possible. *(Hint: one of the equations has no solution)*

a) $2x + 1 = 7$

b) $3x + 4 = 2x + 4$

c) $4x + 1 = 4x + 2$

d) $5x = 7x + 4$

[7]

(MR) **2** Here is a set of equations.

$5x + 1 = 21$ $6x + 1 = x + 21$ $7x + 1 = 2x + 21$ $8x + 1 = 3x + 21$

a) What can you say about the solutions of these equations?

[1]

b) Write down the next equation in the sequence and its solution.

Equation: _____ Solution: $x =$ _____ [2]

3 Solve the equations.

a) $3x + 5 = 21 - x$ b) $\dfrac{4y + 7}{3} = 9$ c) $2(3z - 1) = 4(z + 2)$

$x =$ _____ $y =$ _____ $z =$ _____ [9]

(PS) **4** The diagram shows an isosceles triangle.

The perimeter is 22 cm.

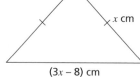

x cm

$(3x - 8)$ cm

Work out the value of x.

$x =$ _____ [3]

Total Marks _____ / 22

5 Solve these equations.

a) $\frac{1}{2}x + 4 = \frac{1}{3}x + 6$

$x =$ _____ [3]

b) $\frac{5x + 3}{6} = \frac{x + 9}{4}$

$x =$ _____ [3]

c) $\frac{4}{x - 1} = \frac{7}{3x - 8}$

$x =$ _____ [3]

(PS) **6** The longest side in this rectangle is 17 cm.

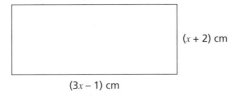

$(x + 2)$ cm

$(3x - 1)$ cm

a) Work out the perimeter.

_____ cm   [4]

b) Work out the area.

_____ cm² [2]

Total Marks _____ / 15

Symmetry and Enlargement

PS **1** Triangles X, Y and Z are mathematically similar.

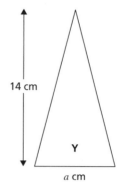

14 cm

7 cm

X

4 cm

Y

a cm

Z

b cm

2 cm

a) Work out the value of *a*.   *a* = ... [2]

b) Work out the value of *b*.   *b* = ... [2]

c) Write the ratio of the perimeters of X to Y to Z.

.... : .... : .... [1]

MR **2** The diagram shows two mathematically similar parallelograms.

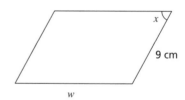

6 cm

40°

7 cm

x

9 cm

w

Decide whether each of the following statements is **true** or **false**. If a statement is false, then correct it.

a) The scale factor of enlargement is +3

.... [1]

b) *w* = 7 × 1.5 = 10.5 cm

.... [1]

c) *x* = 40° × 1.5 = 60°

.... [1]

3 A car is 4.5 metres long. A model of the car is built using a scale of 1 : 18

Work out the length of the model, in centimetres.

.... cm [2]

Total Marks .... / 10

4 Each pair of triangles is congruent. State the condition for congruency in each case.

a)

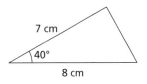

.... [1]

b)

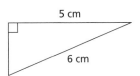 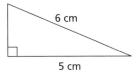

.... [1]

c)

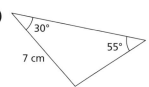 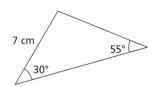

.... [1]

d)

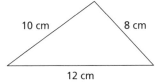

.... [1]

5 A road on a map is 4 cm long. The actual road is 1 km long.

Work out the scale of the map.

.... : .... [2]

6 **a)** Describe fully the transformation from shape
A to shape B.

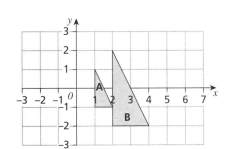

....

.... [3]

b) Describe fully the transformation from shape B
to shape A.

.... [3]

Total Marks .... / 12

Ratio and Proportion

1 150 people visit a park. The ratio of adults to children is 3 : 2

   a) What fraction of the people are children?

   .... [1]

   b) How many adults are there?

   .... [2]

2 The diameter of Earth to the diameter of the Moon is approximately in the ratio 18 : 5

   If the diameter of the Moon is 3500 km, what is the diameter of Earth?

   .... km   [2]

(MR) **3** Matt and Theo go for a walk. For every 4 steps Matt takes, Theo takes 6 steps.

   a) If Matt takes 240 steps, how many would Theo take?

   .... [2]

   b) Natasha says, "If Theo walks 960 steps, Matt will walk 640 steps."

   Is she correct? Show your working.

   .... [2]

(PS) **4** A pencil case contains red pens and blue pens in the ratio 1 : 4
   There are 6 more blue pens than red pens.

   How many blue pens are in the pencil case?

   .... [2]

5 In a flapjack recipe that serves 4 people, there are 250 grams of oats.

   How many grams are needed for a recipe that serves 6 people?

   .... g   [2]

Total Marks .... / 13

(MR) **6** A bus uses 120 litres of fuel to travel 400 miles.

a) How much fuel would the same bus use to travel 280 miles?

.... litres [2]

b) How far would the bus travel on 90 litres of fuel?

.... miles [2]

c) What assumption did you make in working out your answers to part a) and part b)?

.... [1]

(PS) **7** 4 workers take 6 days to build a wall.

a) How long would 5 workers take, assuming they work at the same rate?

.... days [2]

b) If there were 6 workers on the first 2 days and 3 workers for the remaining time, how many days altogether would it take to build the wall?

.... days [3]

8 The cost of a delivery is directly proportional to its mass.
A delivery of a mass of 300 kg costs £45

Another delivery costs £60. What is the mass of this delivery?

.... kg [2]

(MR) **9** Use one of these words to complete each statement.

direct	inverse

a) Speed and time to complete a task are in .... proportion. [1]

b) Speed and distance travelled are in .... proportion. [1]

Total Marks .... / 14

Real-Life Graphs and Rates

1 The distance–time graph shows Tariq's journey one afternoon.

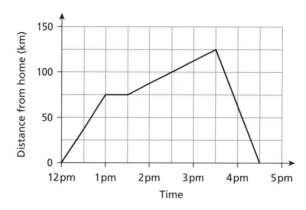

a) How long did Tariq stop for a break?   .... minutes   [1]

b) How far did he travel altogether?

.... km   [2]

c) At what time did Tariq arrive home?   ....   [1]

d) Work out his average speed on the second moving part of the journey.

.... km/h   [2]

(PS) **2** A car travels 190 miles at an average speed of 60 mph.

Did the journey take less than 3 hours? Show your working.

....   [2]

3 A piece of metal has a density of 7.5 g/cm^3. The mass of the metal is 60 grams.

Work out the volume.

.... cm^3 [2]

Total Marks .... / 10

PS **4** A bike is travelling at 20km/h. It slows to half its speed each second.

a) Show this information on the graph.

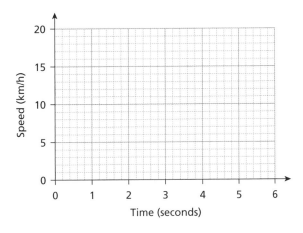

[2]

b) What is the speed of the bike 3 seconds after starting to slow down?

_____ km/h   [1]

MR **5** The number of infections in a small village is shown for 3 days.

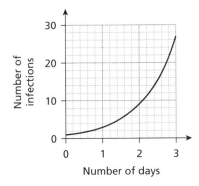

a) Which of these statements are **true**? Circle your answers.

   A The graph shows reciprocal growth.

   B The graph shows exponential growth.

   C The number of infections is growing at the same rate each day.

   D The number of infections is constant each day.   [2]

b) Work out the number of infections after 5 days. State any assumptions you have made.

_____   [3]

Total Marks _____ / 8

Right-Angled Triangles

1 Use Pythagoras' Theorem to work out unknown lengths. Give your answers to 1 decimal place.

a)

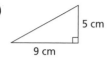

5 cm

9 cm

_____ cm [2]

b)

10 cm

7 cm

_____ cm [2]

2 Decide whether each triangle is right-angled. Show your working.

a)

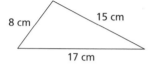

8 cm   15 cm

17 cm

_____

_____ [2]

b)

21 cm

7 cm   20 cm

_____

_____ [2]

c)

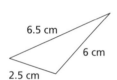

6.5 cm

6 cm

2.5 cm

_____

_____ [2]

3 Work out the perimeter of this triangle.

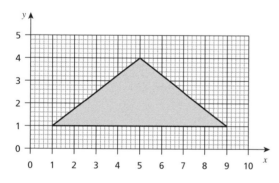

_____ units   [4]

4 Use a calculator to work out the value of the following to 2 decimal places where necessary.

a) 9 tan 20° _____ [1]

b) $\dfrac{12}{\tan 50°}$ _____ [1]

c) $\cos^{-1} 0.5$ _____ [1]

d) $\tan^{-1} 0.466$ _____ [1]

Total Marks _____ / 18

5 Work out the length of each lettered side. You are given that sin 40° = 0.6428

a)

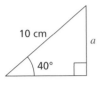

a = ... cm   [2]

b)

b = ... cm   [2]

c)

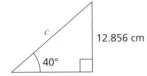

c = ... cm   [2]

6 In each part, work out the size of angle x. Give your answers to 1 decimal place.

a)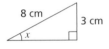

$x =$ .... ° [2]

b)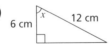

$x =$ .... ° [2]

c)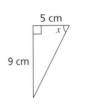

$x =$ .... ° [2]

7 Work out the length x.

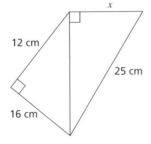

$x =$ .... cm   [4]

PS 8 The diagram shows a section of a mine shaft.

Work out the depth of the shaft, h.

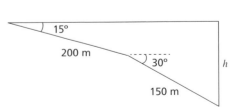

$h =$ .... m   [3]

Total Marks .... / 19

Mixed Test-Style Questions

No Calculator Allowed

1   Here is a Venn diagram.

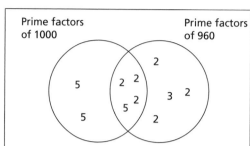

a)   Use the Venn diagram to write 1000 as a product of prime factors.
Give your answer in index form.

....   ☐

2 marks

b)   Use the Venn diagram to work out the highest common factor (HCF) of 1000 and 960

....   ☐

2 marks

2   a)   Use the fact that 14 × 243 = 3402 to work out the answer to 28 × 486

....   ☐

2 marks

b)   Work out 87 × 24 + 87 × 6

....   ☐

2 marks

3 $a - b = -3$, where a and b are integers.

State whether each of the following statements is **true** or **false**.

a) $a > b$

....   ☐

1 mark

b) $a + 3 = b$

....   ☐

1 mark

c) b must be negative

....   ☐

1 mark

4 State whether each of these is **correct** or **incorrect**.

a) $12 \times 109 = 10 \times 100 + 2 \times 9$

1 mark

b) $12 \times 109 = 12 \times 110 - 12 \times 1$

1 mark

c) $12 \times 109 = 10 \times 109 + 2 \times 109$

1 mark

d) $12 \times 109 = 12 \times 100 + 12 \times 9$

1 mark

5 State whether each of these is **correct** or **incorrect**.

a) $6025 \div 25 = (6025 \div 5) \div 5$

1 mark

b) $6025 \div 25 = (6025 \div 20) \div 5$

1 mark

c) $6025 \div 25 = (6025 \div 100) \times 4$

1 mark

6 Jack is charged a booking fee of £4.50 for buying some tickets online. Each ticket costs £20.

Which calculation gives the total cost, in pounds? Circle your answer.

A Number of tickets $\times 24.5$ **B** Number of tickets $\times 90$

C Number of tickets $\times 4.5 + 20$ **D** Number of tickets $\times 20 + 4.5$

1 mark

7 These points lie on a straight line. $(0, -1)$ $(1, 2)$ $(5, 14)$

a) Work out the relationship between the x and y values for this set.

1 mark

b) Work out the coordinates of another point on the same line.

(_____ , _____)

1 mark

Mixed Test-Style Questions

8 Which has the greater answer? $\frac{6}{7} \times \frac{7}{8} \times \frac{8}{9}$ or $\frac{7}{8} \times \frac{8}{9} \times \frac{9}{10}$

Show how you know.

_____

2 marks

9 **a)** On this double number line, 10 and 8 are perfectly aligned.

Write down another pair of numbers that would be aligned.

1 mark

b) On this double number line, 16 and 4 are perfectly aligned.

What number is aligned with 18?

1 mark

10 Work out 438×27

3 marks

11 **a)** Simplify $4x^2 + 8x + 6 - 7x^2 - 13x - 3$

2 marks

b) Expand and simplify $4(3c + 8d) - 5(2c + 7d)$

2 marks

12 Here are some shapes.

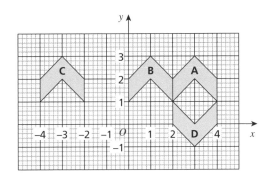

Describe fully the reflections that take:

a) shape A to shape B

_____

1 mark

b) shape A to shape C

_____

1 mark

c) shape A to shape D

_____

1 mark

13 Here is a sequence of patterns with answers.

Pattern 1 $2 + 4 + 4 = 10$

Pattern 2 $2 + 2 + 4 + 4 + 4 = 16$

Pattern 3 $2 + 2 + 2 + 4 + 4 + 4 + 4 = 22$

a) Write down the next row of the sequence.

_____

1 mark

b) Work out the answer to pattern 25

2 marks

c) Which pattern has the answer 298?

2 marks

Mixed Test-Style Questions

14 Here is a number machine.

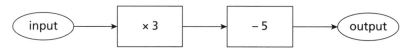

a) Work out the output when the input is 12

1 mark

b) Work out the input when the output is –8

1 mark

15 The diagram shows a triangle with a line drawn parallel to the base.

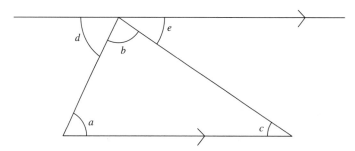

Complete each statement to make it correct.

$d =$ _____ (alternate angles)

$e =$ _____ (alternate angles)

$d + b + e = 180°$ (_____)

so $a + b +$ _____ $= 180°$

Therefore, the angles of a triangle add up to _____ .

5 marks

16 Construct a triangle with sides 3 cm, 6 cm and 8 cm.

2 marks

17 $(x + a)(x + b) = x^2 + ax + bx + 72$, where a and b are positive whole numbers.

a) Write down a pair of possible values for a and b.

$a =$ ....

$b =$ ....

1 mark

b) Write down a different pair of possible values for a and b.

$a =$ ....

$b =$ ....

1 mark

18 Given that $8.3 \times 2.5 = 20.75$, work out:

a) 83×25

....

2 marks

b) 8.3×12.5

....

2 marks

Mixed Test-Style Questions

19 This graph shows the lines $y = \frac{1}{2}x - 1$ and $y = -x + 2$

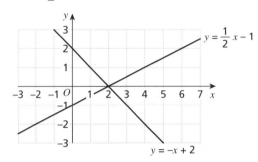

a) Write down the coordinates of a point on the grid where $y > \frac{1}{2}x - 1$ and $y < -x + 2$

(............., ...) [1 mark]

b) Write down the coordinates of a point on the grid where $y > \frac{1}{2}x - 1$ and $y = -x + 2$

(............., ...) [1 mark]

c) Write down the coordinates of the point on the grid where $y = \frac{1}{2}x - 1$ and $y = -x + 2$

(............., ...) [1 mark]

20 Estimate the answer to:

a) 318.1×0.49

[2 marks]

b) $\dfrac{19.4 - 7.2}{2.1}$

[2 marks]

c) $\dfrac{9.6 + 4.31}{0.52}$

[2 marks]

d) $(28.2)^3$

[2 marks]

21 Given that $9x - 8 = 6x + 10$, state whether each equation is **true** or **false**.

a) $9x - 6x = 10 + 8$

.... ☐

1 mark

b) $3x = 18$

.... ☐

1 mark

c) $x = 6$

.... ☐

1 mark

22 Circle the improper fraction that is equivalent to $4\frac{3}{5}$

$\frac{43}{5}$ $\frac{20}{5}$ $\frac{23}{5}$ $\frac{17}{5}$ $\frac{37}{5}$

☐

1 mark

23 The shape on the grid is translated so that point A(1, 1) moves to point A' (2, 5)

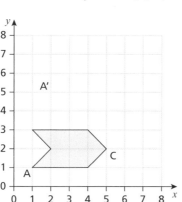

What are the coordinates of the point C' that point C(5, 2) moves to?

(........................, ...) ☐

2 marks

24 A bag of mathematical shapes contains 18 triangles, 26 circles, 4 squares and 2 rectangles. A shape is chosen at random from the bag.

a) What is the probability that the shape is a square?
Give your answer as a fraction in its simplest form.

....

2 marks

b) What is the probability that the shape does **not** have four sides?
Give your answer as a fraction in its simplest form.

....

2 marks

c) One of each shape is now taken from the bag. Another shape is chosen at random.

Azuma says, "Because we have taken out the same number of each shape, the probabilities of choosing each particular shape are unchanged."

Is she correct? Give a reason for your answer.

....

....

1 mark

25 These are the marks for 10 students in a test.

| 8 | 7 | 9 | 2 | 8 | 4 | 8 | 10 | 3 | 6 |

a) Write down the mode.

....

1 mark

b) Work out the median.

....

2 marks

c) Work out the mean.

....

2 marks

d) Work out the range.

....

1 mark

26 The graph shows the attendance at two concerts.

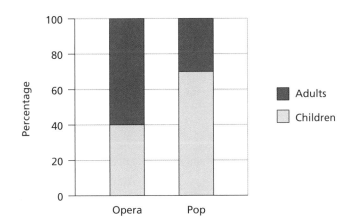

Compare the proportions of adults attending each concert.

....

....

....

1 mark

Total Marks _____ / 84

Mixed Test-Style Questions

Calculator Allowed

1 Show the prime factors of 90 and 144 in the Venn diagram.

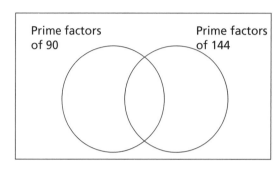

□
3 marks

2 Which of these triangles is **not** a rotation of triangle A?

| A | B | C | D | E |

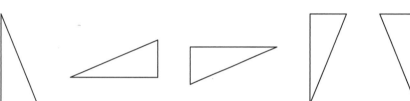

....
□
1 mark

3 Work out the size of angle x.

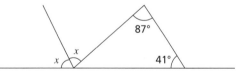

87°

x x

x

41°

.... °
□
3 marks

4 The table shows the times (in seconds) for two students to complete the same task five times.

Amina	12	15	14	11	16
Bob	13	13	14	12	15

a) Give a reason why you think Amina could be the faster.

....

□
1 mark

b) Give a reason why you think Bob could be the faster.

....

□
1 mark

5 Lexi writes $(a + 9)^2 = a^2 + 81$

Ashima writes $(a + 9)^2 = a^2 + 9a + 9a + 81$

Rosie writes $(a + 9)^2 = a^2 + 18a + 81$

Who is correct?

2 marks

6 Here is a sequence. 20, 13, 6, −1, …
Jo thinks the nth term of the sequence is $7n + 13$

Is she correct? Give a reason for your answer.

2 marks

7 The diagram shows an arrowhead.

a) Draw the position of the image of the arrowhead after rotating the object through 90° anticlockwise about A.

2 marks

b) Draw the position of the image of the arrowhead after rotating the object through 90° anticlockwise about C.

2 marks

c) Write down one feature that the images have in common.

1 mark

Mixed Test-Style Questions

8 Here is a way to estimate the adult height for a child.

Add together the height of the mother and father. For a boy add 13 cm and for a girl subtract 13 cm. Then divide the answer by 2.

a) Estimate the adult height of a boy whose parents' heights are 1.74 m and 1.52 m.

.... m

2 marks

b) Estimate the adult height of a girl whose parents' heights are 1.82 m and 1.79 m.

.... m

2 marks

9 Here are some coordinates.　　(−6, −3)　(−1, 2)　(4, 7)

Asha thinks the equation of the line passing through these coordinates is $y = x - 3$

Explain why Asha is **not** correct.

....

1 mark

10 The table shows the favourite food that 30 students eat for breakfast.

Food	Cereal	Toast	Cooked	Fruit	Other/None
Number of students	10	7	3	6	4

a) Complete this table of angle sizes to show the information in a pie chart.

Number of students	30	1	3	4	6	7	10
Angle size	360°						

3 marks

b) Draw the pie chart to represent the information.

Favourite Food

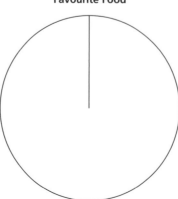

3 marks

11 Decrease £380 by 65%

£ ....

2 marks

12 Solve the equations.

a) $\frac{3}{4}x - 9 = 63$

$x =$ ....

3 marks

b) $\frac{2}{3}(4x - 3) = 18$

$x =$ ....

3 marks

13 Work out the length x. Give your answer to 1 decimal place.

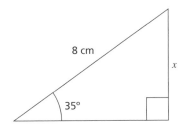

.... cm

2 marks

14 A wooden plank has length 2 metres, width 15 cm and height 4 cm. It has a mass of 7.2 kg.

Work out its density. Give your answer in grams/cm^3

$$\text{density} = \frac{\text{mass}}{\text{volume}}$$

.... g/cm^3

4 marks

15 The diagram shows a rectangle.

Work out the length of the diagonal.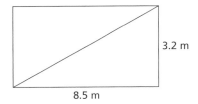

3.2 m

8.5 m

.... m

3 marks

Mixed Test-Style Questions

16 The graph shows the exchange rate between the British pound (£) and the Hong Kong dollar (HK$).

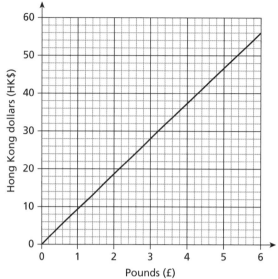

Hong Kong dollars (HK$)

Pounds (£)

a) How many Hong Kong dollars would you get for £900?

HK$ .... ☐

2 marks

b) What is the exchange rate from pounds to Hong Kong dollars?
Give your answer to 2 decimal places.

£1 = HK$ .... ☐

2 marks

17 a) Factorise fully $8x^4 + 12x^2$

.... ☐

2 marks

b) Factorise $x^2 - 5x - 66$

.... ☐

2 marks

18 Three fair coins are thrown.

a) List all the possible outcomes of landing on heads (H) or tails (T).
The first one is done for you.

HHH ....

☐

2 marks

b) What is the probability of landing HHH?

.... ☐

1 mark

19 a) Complete the table of values for the graph of $y = 2x - 3$

x	−2	−1	0	1	2
y					

☐

2 marks

b) Plot the graph of $y = 2x - 3$ for $x = -2$ to $x = 2$

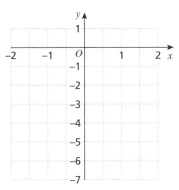

2 marks

c) Write down the coordinates of the y-intercept.

(............,............)

1 mark

d) What is the gradient of the graph of $y = 2x - 3$?

....

1 mark

20 This shape is made from a rectangle and a semicircle.

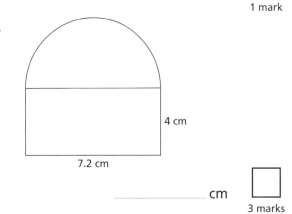

4 cm

7.2 cm

a) Work out the perimeter.
Give your answer to 1 decimal place.

.... cm

3 marks

b) Work out the area.
Give your answer to 1 decimal place.

.... cm^2

3 marks

21 Work out the volume of this triangular prism.

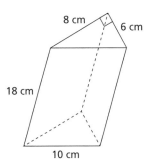

8 cm

6 cm

18 cm

10 cm

.... cm^3

2 marks

22 A jet aircraft has length 36 metres. A model of the jet is made using a scale of 1 : 15

Work out the length of the model in centimetres.

_____ cm ☐ 2 marks

23 A shop sells carpet in 4 m wide rolls or 3 m wide rolls.
Pieces are only cut and sold in full widths.
The carpet costs £25.99 per square metre.

a) What is the cost of buying a piece 5 m long from the 4 m roll?

£ _____ ☐ 2 marks

b) What is the cheaper way to buy a carpet to fit an area measuring 3.7 m by 2.9 m?

_____

_____

☐ 2 marks

24 What is 340 g as a percentage of 2 kg?

_____ % ☐ 2 marks

25 Use your calculator to work out $\frac{848 \times 42}{96}$

Give your answer to 2 significant figures.

_____ ☐ 2 marks

26 Match each description to the correct graph.

| Exponential | Direct proportion | Inverse proportion |

☐ 3 marks

Total Marks _____ / 84

Workbook Answers

Pages 148–149

Number

1. 27 and 25 [2]
   [**1 mark for listing at least three cube numbers (1, 8, 27, …) or at least three square numbers (1, 4, 9, …)**]

2. 1 and 3 [2]
   [**1 mark for one correct answer**]

3. a) Cells ticked as follows:
   80: Multiple of 2, Multiple of 4, Multiple of 5
   81: Multiple of 3
   82: Multiple of 2 [4]
   [**1 mark for each correct column**]
   b) 60 or any multiples of 60, e.g. 120, 180, … [2]
   [**1 mark for $3 \times 4 \times 5$ or $2 \times 3 \times 4 \times 5$ or for listing multiples of 3, 4 and 5**]

4. a) Sometimes true [1]
   b) Always true [1]
   c) Sometimes true [1]

5. 40 seconds [2]
   [**1 mark for listing multiples of 8 and multiples of 10**]

6. $2^2 \times 3$ [1]
   $= 12$ [1]

7. a) 20×4 [1]
   $= 80$ [1]
   b) $4 + 49 - 27$ [1]
   $= 26$ [1]
   c) $10^2 - 8^2 = 100 - 64$ [1]
   $= 36$ [1]

8. $144 = 16 \times 9$ or $2^4 \times 3^2$ [1]
   $x = 4, y = 2$ [1]

9. a) True
   A square number has repeated factors, e.g. $4 = 2 \times 2$ [1]
   b) True
   A prime number has only two factors, itself and 1.
   A cube number has repeated factors, e.g. $3 = 3 \times 1$, $27 = 3 \times 3 \times 3$ [1]
   c) False
   $64 = 8 \times 8$ so is a square number and $64 = 4 \times 4 \times 4$ so is a cube number. [1]

Pages 150–151

Sequences

1. a) A and B [1]
   b) B and D [1]
   c) nth term is $105 - 5n$ or $105 - 5n = 50$ or $55 = 5n$ [1]
   $n = 11$ or 11th term [1]

2. C, D, G and H [2]
   [**1 mark for two correct and no incorrect answers**]

3. nth term is $4n + 1$ or 10th term is $5 + 9 \times 4$ or $5 + 36$ [1]
   $= 41$ (Jack doubled the 5th term) so not correct [1]

4. The term-to-term rule is $+6$ [1]
   The difference between $6n$ and the output in each case is -3, so the nth term is $6n - 3$ [1]

5. a) $n^2 - 1$ [2]
   [**1 mark for n^2**]
   b) 48 and 63 [1]
   One less than a square number or $n = 7$ and $n = 8$ [1]

6. 4, 13, 28, 49 [2]
   [**1 mark for any two correct**]

7. a) $13 + 21 = 34 \, \text{km}$ [1]
   b) $8 \times 3 = 24 \, \text{km}$ [1]
   c) $13 \times 3 = 39$ miles [1]

8. n^3 [1]

Pages 152–153
Perimeter and Area

1. Area of triangle A $= \frac{1}{2} \times 3 \times 2 = 3$ units^2

   Area of triangle B $= \frac{1}{2} \times 3 \times 2 = 3$ units^2

   Area of triangle C $= \frac{1}{2} \times 6 \times 1 = 3$ units^2

   Area of triangle D $= \frac{1}{2} \times 2 \times 3 = 3$ units^2

   Area of triangle E $= \frac{1}{2} \times 3 \times 3 = 4.5$ units^2

   So B, C and D [2]

   [1 mark for any correct area]

2. a) Area of triangle X $= \frac{1}{2} \times a \times h = \frac{1}{2}ah$ [1]

   b) Area of triangle Y $= \frac{1}{2} \times b \times h = \frac{1}{2}bh$ [1]

   c) Area of trapezium $= \frac{1}{2}ah + \frac{1}{2}bh = \frac{1}{2}(a+b)h$ [1]

3. a) 31.4×3 [1]

   $= 94.2$ cm [1]

   b) $(31.4 \div 2) + 10$ or $15.7 + 10$ [1]

   $= 25.7$ cm [1]

4. a) Area $= \pi r^2$, Area $= \pi \times 5^2$ [1]

   $= 78.5$ cm^2 [1]

   b) Area of large square $= 10 \times 10 = 100$ cm^2 or

   Area of right-angled triangle $= \frac{1}{2} \times 5 \times 5 = 12.5$ cm^2 [1]

   Area of small square $= 100 - 4 \times 12.5$ or $4 \times 12.5 = 50$ cm^2 [1]

   Unshaded area $= 78.5 - 50 = 28.5$ cm^2 [1]

5. Area $= \pi r^2$

   Outer area $= \pi \times 45^2$ or $6361.7 \ldots$ [1]

   Inner area $= \pi \times 35^2$ or $3848.4\ldots$ [1]

   Shaded area $= \pi \times 45^2 - \pi \times 35^2$ or $6361.7 \ldots - 3848.4\ldots$ or 2513 cm^2 [1]

   2500 cm^2 [1]

6. a) Area of lawn $= \frac{1}{2} \times (10 + 6) \times 6$ [1]

   $= 48$ m^2 [1]

   Area of shrubs $= \frac{1}{4} \times \pi \times 6^2$ [1]

   $= 9\pi$ cm^2 or 28.3 m^2 [1]

   Area of path $= \frac{1}{2} \times 2 \times 6$ [1]

   $= 6$ m^2 [1]

b) Arc length $= \frac{1}{4} \times \pi \times 12 = 3\pi$ m or 9.42 m [1]

   Perimeter $= 6.3 + 10 + 3\pi + 6 + 2 + 6$ [1]

   $= 39.72$ m or 40 m [1]

Pages 154–155
Statistics and Data

1. Total amount spent $= (4 \times £40) + £60$ or $£160 + £60$ or $£220$ [1]

   Mean amount $= £220 \div 5$ [1]

   $= £44$ [1]

2. a)

	Stall	Circle	Balcony	Total
Adult	240	153	192	585
Child	160	147	108	415
Total	400	300	300	1000

   [2]

   [1 mark for at least two correct values]

   b) Adult tickets – child tickets =

   $585 - 415$ or $(240 - 160) + (153 - 147) + (192 - 108)$ or $80 + 6 + 84$ [1]

   $= 170$ [1]

   c) $585 \times 25 + 415 \times 20$ or $14\,625 + 8300$ [1]

   $= £22\,925$ [1]

3. a) Modal class has the greatest frequency, so 0–9 [1]

   b) Over 200 texts is an outlier, so mean is not suitable. [1]

4. a) $(15 + 12 + 11 + 6 + 6) \div 5$ or $50 \div 5$ [1]

   $= 10$ cm [1]

   b) $6 \times 11 = 66$ [1]

   So extra plant $= 66 - 50 = 16$ cm [1]

5. a) Putting the numbers in order gives

   $x - 3, x - 1, x + 2, x + 4, x + 4$ [1]

   So $x + 2 = 7$, $x = 5$ [1]

   b) Mode is $x + 4$, so $5 + 4 = 9$ [1]

6. a) Greatest possible range $= 18 - 10 = 8$ seconds [1]

   b) Slowest time is $11.6 + 4.9$ [1]

   $= 16.5$ seconds [1]

Pages 156–157

Decimals

1. $5 \times (13 \times 2.6 - 13 \times 0.6) =$
   $5 \times (13 \times (2.6 - 0.6))$ **[1]**
   $= 5 \times 13 \times 2 = 130$ **[1]**
   $10^2 + 10 \times 3 = 100 + 30 = 130$ **[1]**

2. a) 80 **[1]**
   b) 80 **[1]**
   c) 300 **[1]**
   d) 300 **[1]**
   e) 5500 **[1]**

3.

	1 s.f.	2 s.f.	3 s.f.	4 s.f.
1407	1000	1400	1410	1407
2999	3000	3000	3000	2999

[4]

[1 mark for each correct column]

4. a) 8×0.2 **[1]**
   $= 1.6$ **[1]**
   b) $(30 + 9) \div 3$ or $39 \div 3$ **[1]**
   $= 13$ **[1]**
   c) $20 \div 0.5$ or $19 \div 0.5$ **[1]**
   $= 40$ or 38 **[1]**
   d) 40^2 **[1]**
   $= 1600$ **[1]**

5. Perimeter: 40×4 or 39×4 **[1]**
   $= 160\,cm$ or $= 156\,cm$ **[1]**
   Area: 40×40 **[1]**
   $= 1600\,cm^2$ **[1]**

6. a) $12.8 \times 1.5 = 2 \times 6.4 \times 1.5$ or 2×9.6 **[1]**
   $= 19.2$ **[1]**
   b) $6.4 \times 2.5 = 6.4 \times 1.5 + 6.4$ or $9.6 + 6.4$ **[1]**
   $= 16$ **[1]**

7. a) 4700 **[1]**
   b) 831 **[1]**
   c) 0.6 **[1]**
   d) 0.000 135 **[1]**

8. $26\,002\,413 \div 7.7 \times 10^6$ or $3.37\ldots$ **[1]**
   3 (to the nearest whole number) **[1]**

9. $5.678 \times 10^3 \qquad 4.56 \times 10^{-4} \qquad 7.89 \times 10^0$
   $1.234 \times 10^4 \qquad 2.34 \times 10^{-1}$ **[2]**
   [1 mark for any two correct conversions]
   $4.56 \times 10^{-4} \qquad 2.34 \times 10^{-1} \qquad 7.89 \times 10^0$
   $5.678 \times 10^3 \qquad 1.234 \times 10^4$ **[1]**

10. $-0.05 \leqslant \text{error} < 0.05$ **[2]**
    [1 mark for each answer]

Pages 158–159

Algebra

1. There are 24 hours in a day or, for example,
   when $d = 1$, $h = 24$ **[1]**
   $h = 24d$ **[1]**

2. a)

$\times$	$3a$	4
$5a$	$15a^2$	$20a$
4	$12a$	16

[2]

   [1 mark for one correct product]
   b) $15a^2 + 20a + 12a + 16 =$
   $15a^2 + 32a + 16$ **[1]**

3. a) $3x + 24$ **[1]**
   b) $30x + 240$ **[1]**
   c) $3ax + 24a$ **[1]**
   d) $3abx + 24ab$ **[1]**

4. a) $a^2 + 3a + 4a + 12$ **[1]**
   $= a^2 + 7a + 12$ **[1]**
   b) $b^2 + 6b - 2b - 12$ **[1]**
   $= b^2 + 4b - 12$ **[1]**
   c) $c^2 + 5c + 5c + 25$ **[1]**
   $= c^2 + 10c + 25$ **[1]**

5. He has not collected like terms and written
   out the answer as an expression.
   $(3x + 4)(2x + 5y + 6) =$
   $6x^2 + 8x + 15xy + 20y + 18x + 24$
   $= 6x^2 + 26x + 15xy + 20y + 24$ **[1]**

6. $8x^2 - 20x + 2x - 5$ **[1]**
   $= 8x^2 - 18x - 5$ **[1]**

7. $(x^2 + 2x + x + 2)(x + 3)$ or
   $(x^2 + 3x + 2)(x + 3)$ **[1]**
   $= x^3 + 3x^2 + 3x^2 + 9x + 2x + 6$ **[1]**
   $= x^3 + 6x^2 + 11x + 6$ **[1]**

8. a) $PRT = 100I$ [1]

$$PR = \frac{100I}{T}$$ [1]

$$P = \frac{100I}{RT}$$ [1]

9. a) $s = \frac{1}{2}(3 + 7) \times 5$ or $\frac{1}{2} \times 10 \times 5$ [1]

     $= 25$ [1]

b) $9 = \frac{1}{2}(4 + v) \times 6$ or $9 = 3(4 + v)$ [1]

     $3 = 4 + v$ [1]

     $v = -1$ [1]

     Alternatively:

     $2s = (u + v)t$ or $\frac{2s}{t} = u + v$ [1]

     $\frac{2s}{t} - u = v$ or $\frac{2 \times 9}{6} - 4 = v$ or $3 - 4 = v$ [1]

     $v = -1$ [1]

Pages 160–161

3D Shapes: Volume and Surface Area

1. a) 8   [1]

     b) 10   [1]

2. a)

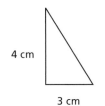

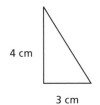

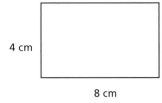

   [4]

[1 mark for both triangular faces; 1 mark for each rectangular face]

b) Area of triangle $\frac{1}{2} \times 3 \times 4$ or $6\,\text{cm}^2$ [1]

     Area of a rectangle is 4×8 or 32 or 5×8 or 40 or 3×8 or 24   [1]

Total surface area =

$(\frac{1}{2} \times 3 \times 4) + (\frac{1}{2} \times 3 \times 4) + (4 \times 8) + (5 \times 8) + (3 \times 8)$ or $6 + 6 + 32 + 40 + 24$ [1]

$= 108\,\text{cm}^2$ [1]

3. a) Volume $= \pi \times 3^2 \times 8$ or 72π or $226.19...$ [1]

     $= 226.2\,\text{cm}^3$ [1]

b) Total surface area =

$\pi \times 3^2 \times 2 + 2 \times \pi \times 3 \times 8$ or $18\pi + 48\pi$ or 66π or $207.34...$ [1]

$= 207.3\,\text{cm}^2$ [1]

c) $25 \div 8 = 3.125$, so 3   [1]

4. a) Area of triangle $= \frac{1}{2} \times 8 \times 3$ or $12\,\text{cm}^2$ [1]

     Volume of prism $= \frac{1}{2} \times 8 \times 3 \times 12$ or 12×12 [1]

     $= 144\,\text{cm}^3$ [1]

b) Total surface area =

$(\frac{1}{2} \times 8 \times 3) + (\frac{1}{2} \times 8 \times 3) + (8 \times 12) + (5 \times 12) + (5 \times 12)$ or $12 + 12 + 96 + 60 + 60$ [1]

$= 240\,\text{cm}^2$ [1]

5. a) Cuboid has 6 faces ($F = 6$), 8 vertices ($V = 8$) and 12 edges ($E = 12$)   [1]

     $6 + 8 = 12 + 2 (= 14)$ [1]

b) Triangular prism has 5 faces ($F = 5$), 6 vertices ($V = 6$) and 9 edges ($E = 9$)   [1]

     $5 + 6 = 9 + 2 (= 11)$ [1]

c) Square-based pyramid has 5 faces ($F = 5$), 5 vertices ($V = 5$) and 8 edges ($E = 8$)   [1]

     $5 + 5 = 8 + 2 (= 10)$ [1]

6. Area of cross-section $= (12 \times 8) + (4 \times 2)$ or $96 + 8$ or $104\,\text{cm}^2$ [1]

     Volume $= 104 \times 5$ [1]

     $= 520\,\text{cm}^3$ [1]

Pages 162–163

Interpreting Data

1. a) Year 8   [1]

     b) There are more than 150 students in Year 11   [1]

2. $360°$ represents 24 students, so $360° \div 24 = 15°$ represents 1 student **[1]**

So $45° \div 15° = 3$ students get 0 homework
$60° \div 15° = 4$ students get 1 homework
$90° \div 15° = 6$ students get 2 homeworks
$165° \div 15° = 11$ students get 3 homeworks **[1]**

Total number of homeworks $= 0 + (4 \times 1) +$
$(6 \times 2) + (11 \times 3) = 4 + 12 + 33 = 49$ **[1]**
Mean number of homeworks =
$49 \div 24 = 2.04$ **[1]**

3. Range for farm A is $9 - 5 = 4$ years
Range for farm B is $11 - 4 = 7$ years **[1]**
This appears to support Ted but 4 is an outlier and does not seem generally representative of the animals on farm B, so without that single value the range for farm B would be $11 - 8 = 3$ years, so farm B looks to have more consistent ages. **[1]**

4. a)

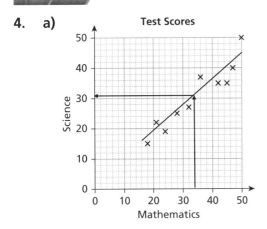

[3]

[1 mark for five correctly plotted points;
1 mark for line of best fit drawn through the points]

b) Reading from line of best fit, estimated score is 31 (answer may vary slightly depending on line of best fit drawn) **[1]**

5. a) Car B, as the graph goes lower **[1]**

b) Car A (it lost £15 000 of its value in year 1 and £15 000 is 30% of £50 000) **[1]**

c) $\dfrac{10000}{50000}$ or $\dfrac{1}{5}$ **[1]**

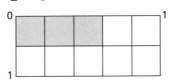

Pages 164–165

Fractions

1. $\dfrac{1}{2} \times \dfrac{3}{5} = \dfrac{3}{10}$

[2]

[1 mark for correct shading of rectangle]

2. a) $5 \times \dfrac{1}{3} = \dfrac{5}{3}$ **[1]**

b) $\dfrac{1}{3} \times 4 = \dfrac{4}{3}$ and $1\dfrac{2}{3} = \dfrac{5}{3}$

So $\dfrac{1}{3} \times 4 < 1\dfrac{2}{3}$ **[1]**

c) $\dfrac{1}{2} \times \dfrac{1}{3} = \dfrac{1}{6}$

So $\dfrac{1}{5} > \dfrac{1}{2} \times \dfrac{1}{3}$ **[1]**

d) $\dfrac{1}{3} \times \dfrac{1}{2} = \dfrac{1}{6}$

So $\dfrac{1}{3} \times \dfrac{1}{2} < \dfrac{2}{5}$ **[1]**

3. a) $\dfrac{1}{12}$ **[1]**

b) $\dfrac{5}{12}$ **[1]**

c) $\dfrac{1}{2}$ **[1]**

d) $\dfrac{2}{5}$ **[1]**

4. a) $\dfrac{14}{9} \times 234 = 2 \times \dfrac{7}{9} \times 234$ or 2×182 **[1]**

$= 364$ **[1]**

b) $\dfrac{7}{3} \times 234 = 3 \times \dfrac{7}{9} \times 234$ or 3×182 **[1]**

$= 546$ **[1]**

c) $\dfrac{7}{18} \times 234 = \dfrac{1}{2} \times \dfrac{7}{9} \times 234$ or $\dfrac{1}{2} \times 182$ **[1]**

$= 91$ **[1]**

5. $2\dfrac{1}{2} - \dfrac{3}{4}$ or 2 hours 30 minutes – 45 minutes

$= 1$ hour 45 minutes **[1]**

$= 1\dfrac{3}{4}$ hours **[1]**

6. a) $1\dfrac{1}{3} + 2\dfrac{3}{4} + 1\dfrac{1}{2} = 1\dfrac{4}{12} + 2\dfrac{9}{12} + 1\dfrac{6}{12}$ **[1]**

$4 + \dfrac{19}{12}$ or $4 + 1\dfrac{7}{12}$ **[1]**

$= 5\dfrac{7}{12}$ **[1]**

b) $1\frac{5}{8} + 7\frac{2}{3} = 11\frac{7}{24} - 3 + 1 = 9\frac{7}{24}$ **[1]**

c) $11\frac{7}{24} - 8\frac{2}{3} = 4\frac{5}{8} - 2 = 2\frac{5}{8}$ **[1]**

7. a) $1\frac{1}{2}$ **[1]**

b) $4\frac{1}{2} \times 1\frac{1}{2} = \frac{9}{2} \times \frac{3}{2}$ **[1]**

$= \frac{27}{4}$ or $6\frac{3}{4}$ **[1]**

8. a) Aisha **[1]**

b) $7 + 1\frac{3}{4}$ **[1]**

$= 8\frac{3}{4}$ **[1]**

Pages 166–167

Coordinates and Graphs

1. a) $y = 2x + 3$ **[2]**

[1 mark for finding a correct relationship for one point, e.g. $y = x + 7$ for (4, 11)]

b) Yes, $y = 2x + 3$ gives a straight line **[1]**

2. D **[1]**

3. Correct equation is $x = y - 6$ **[2]**

[1 mark for checking the coordinates of a point, e.g. when $x = -4$, $y = -10$]

4. a)

x	–3	–2	–1	0	1	2	3
y	7	2	–1	–2	–1	2	7

[2]

[1 mark for at least two correct values]

b)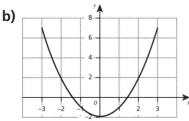

[2]

[1 mark for at least three correct points plotted]

5. a)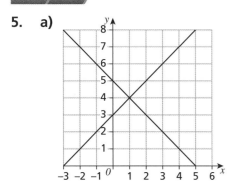

[2]

[1 mark for each line]

b) $x = 1$, $y = 4$ **[1]**

6.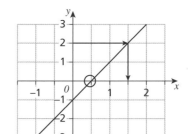

a) $x = \frac{1}{2}$ **[1]**

b) $x = 1\frac{1}{2}$ **[1]**

Pages 168–169

Angles

1. $b = 180° - 35° = 145°$ (angles on a straight line add up to 180°/ are supplementary) **[1]**

$c = 145°$ (alternate angles) **[1]**

2. $a = 180° - 100° = 80°$ **[1]**

$b = 45°$ **[1]**

$c = 100°$ **[1]**

$d = 180° - 45° = 135°$ **[1]**

$e = 135°$ **[1]**

$f = 180° - 80° - 45° = 55°$ **[1]**

3. a) Rhombus (four equal sides) **[1]**

b) Kite (adjacent pairs of equal sides) **[1]**

c) Arrowhead or delta (adjacent pairs of equal sides and a reflex angle) **[1]**

4. Octagon split into 6 triangles so sum of interior angles = $6 \times 180°$ **[1]**

$= 1080°$ **[1]**

5. $a = 360° \div 5 = 72°$ **[1]**

Exterior angle of a regular polygon = $360° \div$ number of sides **[1]**

$b = 72° \div 2 = 36°$ **[1]**

Exterior angle of isosceles triangle = sum of interior opposite angles **[1]**

$c = 108° - 36° = 72°$ **[1]**

Interior angle of pentagon = $180° - 72° = 108°$ **[1]**

6. Sum of interior angles of a hexagon = 720°

[1]

$(720 - (4 \times 145)) \div 2 = 70°$ [1]

7. $360° \div (180° - 135°)$ or $360° \div 45°$ [1]

= 8 sides [1]

8.

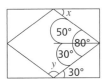

$x = 50°$ [1]

$y = 180° - 80° = 100°$ [1]

Pages 170–171

Probability

1. a) $\frac{1}{6}$ [1]

b) $\frac{2}{6}$ or $\frac{1}{3}$ [1]

c) $\frac{3}{6}$ or $\frac{1}{2}$ [1]

d) $\frac{1}{2} \times 50 = 25$ [1]

2. a) A = {3, 6, 9, 12, 15, 18, 21} [1]

b) B´ = {12, 15, 21} [1]

c) A ∪ B = {1, 2, 3, 6, 9, 12, 15, 18, 21} [1]

d) Factors of 18 [1]

e) $\frac{4}{9}$ [1]

3. a) (HH), HT, TH, TT [1]

b) $\frac{1}{4}$ [1]

4. a) $\frac{12}{30}$ [1]

$= \frac{2}{5}$ [1]

b) $1 - \frac{2}{5}$ [1]

$= \frac{3}{5}$ [1]

5. a) $\frac{100}{300}$ [1]

$= \frac{1}{3}$ [1]

b) $\frac{300 - 80}{300}$ or $\frac{220}{300}$ [1]

$= \frac{11}{15}$ [1]

c) $120 \div 3 \times 2 = 80$ yellow bricks [1]

$\frac{80}{300}$ [1]

$= \frac{4}{15}$ [1]

d) $\frac{90}{260}$ [1]

$= \frac{9}{26}$ [1]

Pages 172–173

Fractions, Decimals and Percentages

1. a) 10% of £40 = £4 or £40 + 3 × £4 or

1.3 × £40 [1]

= £52 [1]

b) 10% of £30 = £3 or £30 + 4 × £3 or

1.4 × £30 [1]

= £42 [1]

c) 10% of £50 = £5 or £50 − 2 × £5 or

0.8 × £50 [1]

= £40 [1]

d) 10% of £60 = £6 or £60 − 7 × £6 or

0.3 × £60 [1]

= £18 [1]

2. a) Boxes joined as follows:

Increase £40 by 15% to £40 × 1.15 [1]

Increase £15 by 40% to £15 × 1.4 [1]

Decrease £40 by 15% to £40 × 0.85 [1]

Decrease £15 by 40% to £15 × 0.6 [1]

b) £40 × 1.5   Increase £40 by 50% [1]

£40 × 0.6   Decrease £40 by 40% [1]

3. $\frac{15}{25} \times 100$ or 15 × 4 [1]

= 60% [1]

4. a) 1% of £500 = £5 or 4% of £500 =

$\frac{4}{100} \times 500$ [1]

= £20 [1]

b) £500 + 3 × £20 [1]

= £560 [1]

5. Percentage reduction = $\frac{45}{75} \times 100\%$ [1]

= 60% [1]

6. **a)** She is taxed on £45 000 – £12 500 =

£32 500 [1]

20% of £32 500 = 0.2 × £32 500 [1]

= £6500 [1]

b) Pay = £45 000 – £6500 = £38 500 per year

[1]

= £38 500 ÷ 12 or £3208.33 [1]

= £3200 (to the nearest £100) [1]

7.

Energy used	250 kWh @ 28p per kWh	£70.00
Standing charge	30 days @ 48p per day	£14.40
Subtotal of charges before VAT		£84.40
VAT @ 5%		£4.22
Total electricity charges		£88.62

[5]

[1 mark for each correct row]

Pages 174–175

Equations

1. **a)** $2x = 7 - 1$ or $2x = 6$ [1]

$x = 3$ [1]

b) $3x - 2x = 4 - 4$ [1]

$x = 0$ [1]

c) No solution [1]

d) $-4 = 7x - 5x$ or $-4 = 2x$ [1]

$x = -2$ [1]

2. **a)** All have same solution [1]

b) $9x + 1 = 4x + 21$ [1]

So $5x + 1 = 21$, $5x = 20$, $x = 4$ [1]

3. **a)** $3x + x = 21 - 5$ [1]

$4x = 16$ [1]

$x = 4$ [1]

b) $4y + 7 = 9 \times 3$, $4y + 7 = 27$ [1]

$4y = 27 - 7$, $4y = 20$ [1]

$y = 5$ [1]

c) $6z - 2 = 4z + 8$ [1]

$6z - 4z = 8 + 2$, $2z = 10$ [1]

$z = 5$ [1]

4. $3x - 8 + x + x = 22$ [1]

$5x - 8 = 22$, $5x = 22 + 8$, $5x = 30$ [1]

$x = 6$ [1]

5. **a)** $\frac{1}{2}x - \frac{1}{3}x = 6 - 4$ or $3x + 24 = 2x + 36$ [1]

$\frac{3}{6}x - \frac{2}{6}x = 2$ or $\frac{1}{6}x = 2$ or $3x - 2x = 36 - 24$ [1]

$x = 12$ [1]

b) $4(5x + 3) = 6(x + 9)$ or $20x + 12 = 6x + 54$ [1]

$20x - 6x = 54 - 12$ or $14x = 42$ [1]

$x = 3$ [1]

c) $4(3x - 8) = 7(x - 1)$ or $12x - 32 = 7x - 7$ [1]

$12x - 7x = 32 - 7$ or $5x = 25$ [1]

$x = 5$ [1]

6. **a)** $3x - 1 = 17$ [1]

$3x = 18$, $x = 6$ [1]

Width = $6 + 2 = 8\,cm$ [1]

Perimeter = $17 + 8 + 17 + 8 = 50$ cm [1]

b) Area = 17×8 [1]

= $136\,cm^2$ [1]

Pages 176–177

Symmetry and Enlargement

1. **a)** Scale factor = $14 \div 7 = 2$ [1]

$a = 4 \times 2 = 8$ [1]

b) Scale factor = $2 \div 4 = 0.5$ or $4 \div 2 = 2$ [1]

$b = 7 \times 0.5$ or $7 \div 2 = 3.5$ [1]

c) Ratio of perimeters = $4 : 8 : 2 = 2 : 4 : 1$ [1]

2. **a)** False, the scale factor of enlargement

is $9 \div 6 = 1.5$ [1]

b) True [1]

c) False, $x = 40°$ as angles are preserved

during enlargements [1]

3. $4.5\,m = 450\,cm$

$450 \div 18$ [1]

= $25\,cm$ [1]

4. **a)** SAS [1]

b) SSS [1]

c) RHS [1]

d) AA corr S [1]

5. 4 cm : 1000 m or 4 cm : 100 000 cm [1]
   1 : 25 000 [1]
6. a) Enlargement [1]
   Centre (0, 0) [1]
   Scale factor 2 [1]
   b) Enlargement [1]
   Centre (0, 0) [1]
   Scale factor $\frac{1}{2}$ [1]

Pages 178–179
Ratio and Proportion

1. a) $\frac{2}{5}$ [1]
   b) $150 \div 5 \times 3$ [1]
   $= 90$ adults [1]
2. $3500 \div 5 \times 18$ [1]
   $= 12\,600$ km [1]
3. a) Ratio of steps is 2 : 3, $240 \times \frac{3}{2}$ [1]
   $= 360$ [1]
   b) $960 \times \frac{2}{3}$ or $640 \times \frac{3}{2}$ [1]
   $960 \times \frac{2}{3} = 640$ or $640 \times \frac{3}{2} = 960$,
   so correct [1]
4. 3 parts = 6 pens [1]
   So 1 part is 2 pens and 4 parts is 8 pens,
   so there are 8 blue pens [1]
5. $250 \div 4 \times 6$ [1]
   $= 375$ g [1]

6. a) $120 \div 400 \times 280$ [1]
   $= 84$ litres [1]
   b) $400 \div 120 \times 90$ [1]
   $= 300$ miles [1]
   c) Same rate of consumption [1]
7. a) $4 \times 6 = 24$ worker days [1]
   $24 \div 5 = 4.8$ days [1]
   b) $6 \times 2 = 12$, so 12 worker days left [1]
   $12 \div 3 = 4$ days [1]
   6 days altogether [1]
8. $300 \div 45 \times 60$ [1]
   $= 400$ kg [1]
9. a) inverse [1]
   b) direct [1]

Pages 180–181
Real-Life Graphs and Rates

1. a) 30 minutes [1]
   b) $125 \times 2 = 250$ km [2]
   [1 mark for 125]
   c) 4.30 pm [1]
   d) $(125 - 75) \div 2$ or $50 \div 2$ [1]
   $= 25$ km/h [1]
2. $190 \div 60$ [1]
   $= 3.16\ldots$ so over 3 hours [1]
   Alternatively:
   60×3 [1]
   $= 180$ miles so 10 miles remaining after
   3 hours [1]
3. Volume $= \frac{\text{mass}}{\text{density}} = \frac{60}{7.5}$ [1]
   $= 8$ grams [1]

4. a)

[2]
   b) 2.5 km/h [1]
5. a) B [1] and C [1]
   b) $27 \times 3 \times 3$ [1]
   $= 243$ [1]
   Assume continues at same rate (scale
   factor 3 each day) [1]

Pages 182–183
Right-Angled Triangles

1. a) $9^2 + 5^2 = 81 + 25 = 106$ [1]
   $\sqrt{106} = 10.3$ cm (1 d.p.) [1]
   b) $10^2 - 7^2 = 100 - 49 = 51$ [1]
   $\sqrt{51} = 7.1$ cm (1 d.p.) [1]
2. a) $8^2 + 15^2 = 64 + 225 = 289$ [1]
   $= 17^2$, so right-angled [1]
   b) $7^2 + 20^2 = 49 + 400 = 449$ [1]
   $\neq 21^2$, so not right-angled [1]

c) $2.5^2 + 6^2 = 6.25 + 36 = 42.25$ [1]

$= 6.5^2$, so right-angled [1]

3. Using Pythagoras' Theorem $l^2 = 4^2 + 3^2$,

where l is the slant height. [1]

$l^2 = 16 + 9$, $l^2 = 25$ [1]

$l = 5$ units [1]

Perimeter $= 8 + 5 + 5 = 18$ units [1]

4. a) 3.28 [1]

b) 10.07 [1]

c) 60° [1]

d) 25° (or 24.99°) [1]

5. a) $a = 10 \sin 40°$ [1]

$= 6.428$ cm [1]

b) $b = 5 \sin 40°$ [1]

$= 3.214$ cm [1]

c) $\sin 40° = \dfrac{12.856}{c}$ [1]

$c = \dfrac{12.856}{0.6428} = 20$ cm [1]

6. a) $\sin x = \dfrac{3}{8}$ [1]

$x = \sin^{-1}(\dfrac{3}{8}) = 22.0°$ [1]

b) $\cos x = \dfrac{6}{12}$ [1]

$x = \cos^{-1}(\dfrac{6}{12}) = 60.0°$ [1]

c) $\tan x = \dfrac{9}{5}$ [1]

$x = \tan^{-1}(\dfrac{9}{5}) = 60.9°$ [1]

7. $12^2 + 16^2 = 144 + 256$ [1]

$= 400$ or $\sqrt{400} = 20$ [1]

$25^2 - 400$ or $25^2 - 20^2 = 625 - 400 = 225$ [1]

$x = 15$ cm [1]

8. $200 \sin 15°$ or $150 \sin 30°$ [1]

$200 \sin 15° + 150 \sin 30°$ or $51.76 \ldots + 75$ [1]

126.76 or 126.8 [1]

Mixed Test-Style Questions

Pages 184–193

No Calculator Allowed

1. a) $1000 = 2^3 \times 5^3$ [2]

[1 mark for choosing 2, 2, 2, 5, 5, 5]

b) HCF $= 2^3 \times 5$ [1]

$= 40$ [1]

2. a) $28 \times 486 = 2 \times 14 \times 2 \times 243$ or

$2 \times 2 \times 3402$ [1]

$= 13\,608$ [1]

b) $87 \times 24 + 87 \times 6 = 87 \times 30$ [1]

$= 2610$ [1]

3. a) False [1]

b) True [1]

c) False [1]

4. a) Incorrect [1]

b) Correct [1]

c) Correct [1]

d) Correct [1]

5. a) Correct [1]

b) Incorrect [1]

c) Correct [1]

6. D Number of tickets $\times 20 + 4.5$ [1]

7. a) $y = 3x - 1$ [1]

b) Any coordinates satisfying the equation

$y = 3x - 1$, e.g (2, 5) [1]

8. $\dfrac{6}{7} \times \dfrac{7}{8} \times \dfrac{8}{9} = \dfrac{6}{9}$ (0.666…) and $\dfrac{7}{8} \times \dfrac{8}{9} \times \dfrac{9}{10}$

$= \dfrac{7}{10}$ (0.7)

So the second calculation gives the greater

answer. [2]

[1 mark for one correct calculation or

conversion]

9. a) Any numbers in the ratio 5 : 4,

e.g. 5 and 4 or 20 and 16 [1]

b) $18 \div 4 = 4.5$ [1]

10. 11 826 [3]

[2 marks for 3066 and 8760 or 10 800 and

810 and 216; 1 mark for any two of 8000,

2800, 600, 210, 160, 56 or any one of 3066,

8760, 10 800, 810 or 216]

11. a) $-3x^2 - 5x + 3$ [2]

[1 mark for two correct terms]

b) $12c + 32d - 10c - 35d = 2c - 3d$ [2]

[1 mark for correct expansion or $2c$ or $-3d$]

12. a) Reflection in the line $x = 2$ [1]

b) Reflection in the line $x = 0$ or y-axis [1]

c) Reflection in the line $y = 1$ [1]

13. a) $2 + 2 + 2 + 2 + 4 + 4 + 4 + 4 + 4 = 28$ [1]

b) $25 \times 2 + 26 \times 4$ [1]

$= 154$ [1]

c) $(298 - 4) \div (2 + 4)$ or

   $(298 - 4) \div 6$ [1]

   $= 49$, so pattern 49 [1]

14. a) $12 \times 3 - 5 = 31$ [1]

  b) $(-8 + 5) \div 3 = -1$ [1]

15. $d = a$ (alternate angles) [1]

  $e = c$ (alternate angles) [1]

  $d + b + e = 180°$ (angles on a straight line

  are supplementary (or add up to 180°)) [1]

  so $a + b + c = 180°$ [1]

  Therefore, the angles of a triangle add up

  to 180° [1]

16. Example answer: using 8 cm as the base line,
arc from one end of radius 6 cm, arc from
other end of radius 3 cm, triangle completed
through point of intersection of the arcs.

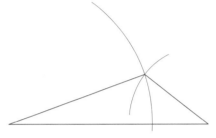

[2]

**[1 mark for a correct baseline and one
correct arc]**

Alternative diagram using full circles:

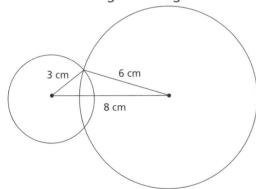

17. a) Any pair from: 1 and 72 or 2 and 36 or
3 and 24 or 4 and 18 or 6 and 12 or
8 and 9 [1]

  b) A different pair from the list above [1]

18. a) $83 \times 25 = 8.3 \times 10 \times 2.5 \times 10$

   $= 20.75 \times 10 \times 10$ [1]

   $= 2075$ [1]

  b) $8.3 \times 12.5 = 8.3 \times 2.5 + 8.3 \times 10$

   $= 20.75 + 83$ or $8.3 \times 2.5 \times 5 = 20.75 \times 5$ [1]

   $= 103.75$ [1]

19. a) Any point in the region marked A,
e.g. (0, 1) [1]

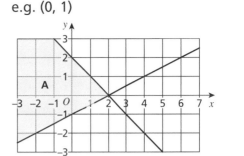

  b) Example answers: $(-1, 3)$, $(0, 2)$, $(1, 1)$ [1]

  c) (2, 0) [1]

20. a) 300×0.5 or 320×0.5 [1]

   $= 150$ or $= 160$ [1]

  b) $(20 - 7) \div 2$ or $(19 - 7) \div 2$ [1]

   $= 6.5$ or $= 6$ [1]

  c) $(10 + 4) \div 0.5$ [1]

   $= 28$ [1]

  d) 30^3 [1]

   $= 27\,000$ [1]

21. a) True [1]

  b) True [1]

  c) True [1]

22. $\dfrac{23}{5}$ [1]

23. (6, 6) [2]

**[1 mark for one correct coordinate or for
drawing the correct translation]**

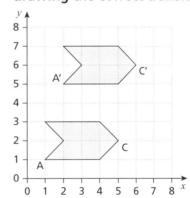

24. a) $\dfrac{4}{50}$ [1]

   $= \dfrac{2}{25}$ [1]

  b) $\dfrac{44}{50}$ [1]

   $= \dfrac{22}{25}$ [1]

  c) Not correct as for example there are
46 left, so now P(square) $= \dfrac{3}{46} \neq \dfrac{2}{25}$ [1]

25. **a)** 8 [1]

b) Putting numbers in order

2, 3, 4, 6, 7, 8, 8, 8, 9, 10 [1]

Median = 7.5 [1]

c) (2 + 3 + 4 + 6 + 7 + 8 + 8 + 8 + 9 + 10) ÷ 10

or 65 ÷ 10 [1]

= 6.5 [1]

d) 10 − 2 = 8 [1]

26. A greater proportion of adults attended the opera (60% compared with 30%) [1]

Pages 194–200

Calculator Allowed

1.

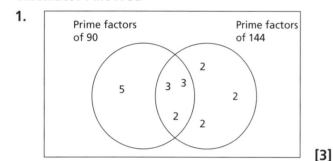

[3]

[1 mark for each correct region]

2. D [1]

3. 87° + 41° or 128° or 180 − 87° − 41° or 52° [1]

(180° − 52°) ÷ 2 or 128° ÷ 2 [1]

$x = 64°$ [1]

4. **a)** Any suitable answer, e.g. Amina had the shortest time, 11 seconds. [1]

b) Any suitable answer, e.g. Bob had the lowest total time 67 seconds (total for Amina was 68 seconds) [1]

5. Ashima and Rosie are both correct but Ashima has not simplified the answer. [2]

[1 mark if only one chosen]

6. Not correct as sequence decreases by 7 each time or $7n$ would mean sequence is increasing **[1]** So nth term $= -7n + 27$ or $7n + 13$ gives 20, 27, 34, 41, ... **[1]**

7. **a)**

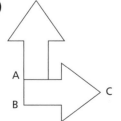

[2]

[1 mark for correct orientation]

b)

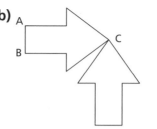

[2]

[1 mark for correct orientation]

c) Both images have the same orientation (point the same way) [1]

8. **a)** 1.74 + 1.52 + 0.13 or 3.39 [1]

3.39 ÷ 2 = 1.695 m [1]

b) 1.82 + 1.79 − 0.13 = 3.48 [1]

3.48 ÷ 2 = 1.74 m [1]

9. Each y-coordinate is 3 more than its x-coordinate, so equation is $y = x + 3$ [1]

10. **a)**

Number of students	30	1	3	4	6	7	10
Angle size	360°	12°	36°	48°	72°	84°	120°

[3]

[2 marks for four correct angles, 1 mark for two correct angles]

b)

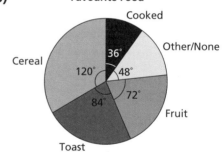

Favourite Food

Cooked	Other/ None	Fruit	Toast	Cereal
36°	48°	72°	84°	120°

[3]

[2 marks for at least three correct sectors; 1 mark for at least one correct sector]

11. 100% − 65% = 35%, so multiplier is 0.35 or 0.35 × £380 [1]

= £133 [1]

12. **a)** $\frac{3}{4}x = 63 + 9$ or $\frac{3}{4}x = 72$ [1]

$3x = 4 × 72$ or $\frac{1}{4}x = 72 ÷ 3$ or $3x = 288$

or $\frac{1}{4}x = 24$ [1]

$x = 96$ [1]

b) $2(4x - 3) = 54$ **[1]**

$8x - 6 = 54$ or $8x = 60$ or $4x - 3 = 27$ or

$4x = 30$ **[1]**

$x = 7.5$ **[1]**

13. $\sin 35° = \frac{x}{8}$ or $x = 8\sin 35°$ **[1]**

$= 4.588... = 4.6\,cm$ (to 1 d.p.) **[1]**

14. Volume is $200 \times 15 \times 4 = 12\,000\,cm^3$ **[1]**

Mass $= 7200\,g$ **[1]**

Density $= 7200 \div 12\,000$ **[1]**

$= 0.6\,g/cm^3$ **[1]**

15. $(\text{diagonal}^2 =)\ 8.5^2 + 3.2^2$ or 82.49 **[1]**

diagonal $= \sqrt{8.5^2 + 3.2^2}$ or $\sqrt{82.49}$ **[1]**

$= 9.08\,m$ or $9.1\,m$ **[1]**

16. a) $28 \times 3 \times 100$ or 28×300 **[1]**

$= HK\$ 8400$ **[1]**

b) $£1 = 28 \div 3$ **[1]**

$= HK\$ 9.33$ **[1]**

17. a) $4x^2(2x^2 + 3)$ **[2]**

[1 mark for partial factorisation, $4x(2x^3 + 3x)$ or $x^2(8x^2 + 12)$ or $4(2x^4 + 3x^2)$ or $x(8x^3 + 12x)$]

b) $(x - 11)(x + 6)$ **[2]**

[1 mark for $(x + a)(x + b)$ where $ab = -66$ or $a + b = -5$]

18. a) HHH, HHT, HTH, THH, HTT, THT, TTH,

TTT **[2]**

[1 mark for at least five correct]

b) $\frac{1}{8}$ **[1]**

19. a)

x	-2	-1	0	1	2
y	-7	-5	-3	-1	1

[2]

[1 mark for at least two correct values]

b)

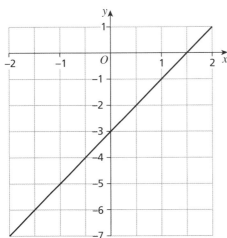

[2]

[1 mark for at least two correct points plotted]

c) $(0, -3)$ **[1]**

d) 2 **[1]**

20. a) Circumference of circle $= \pi \times 7.2$ or

Circumference of semicircle $=$

$\pi \times 7.2 \div 2$ **[1]** $= 11.3...\,cm$ **[1]**

Perimeter $= 11.3 + 4 + 7.2 + 4 = 26.5\,cm$

(1 d.p.) **[1]**

b) Area of rectangle is $7.2 \times 4 = 28.8\,cm^2$ **[1]**

Area of circle is $\pi \times 3.6^2$ or Area of

semicircle is $\pi \times 3.6^2 \div 2 = 20.357...$ **[1]**

Area of shape is $28.8 + 20.357...$

$= 49.2\,cm^2$ (to 1 d.p.) **[1]**

21. Area of cross-section is $\frac{1}{2} \times 8 \times 6 = 24\,cm^2$ **[1]**

Volume of prism $= \frac{1}{2} \times 8 \times 6 \times 18$ or 24×18

$= 432\,cm^3$ **[1]**

22. $36 \div 15 = 2.4\,m$ or $3600 \div 15$ **[1]**

$= 240\,cm$ **[1]**

23. a) $5 \times 4 \times £25.99$ **[1]**

$= £519.80$ **[1]**

b) Buying from the 3 m wide roll $3.7 \times 3 =$

$11.1\,m^2$ or buying from the 4 m wide roll

$2.9 \times 4 = 11.6\,m^2$ **[1]**

Buying from the 3 m wide roll needs less

carpet **[1]**

24. $\frac{340}{2000} \times 100\%$ **[1]**

$= 17\%$ **[1]**

25. 371 **[1]**

$= 370$ (to 2 s.f.) **[1]**

26.

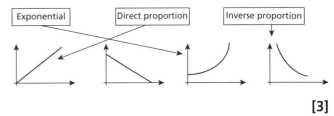

[3]

[1 mark for each correct match]

Graph Paper

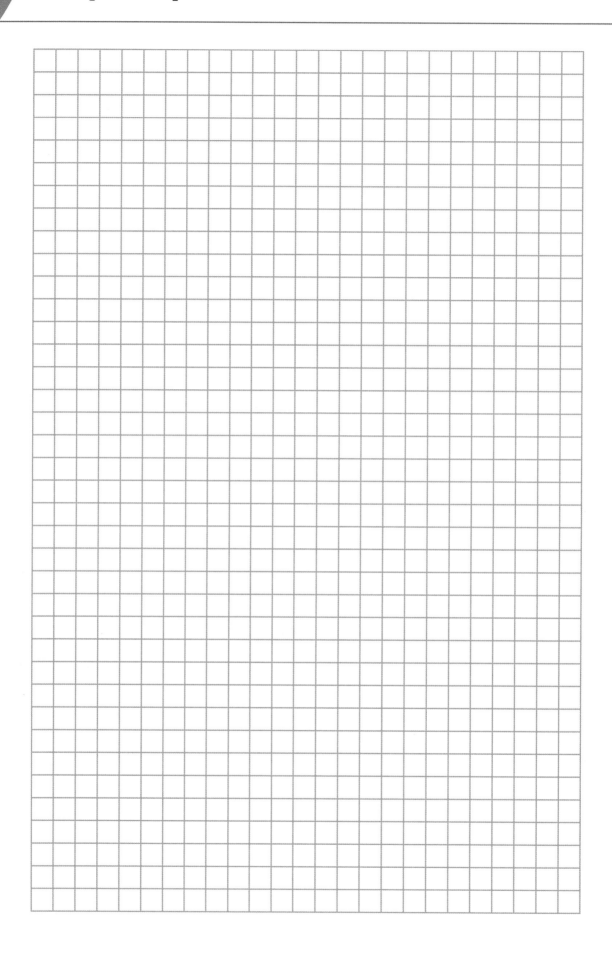

KS3 Maths Workbook

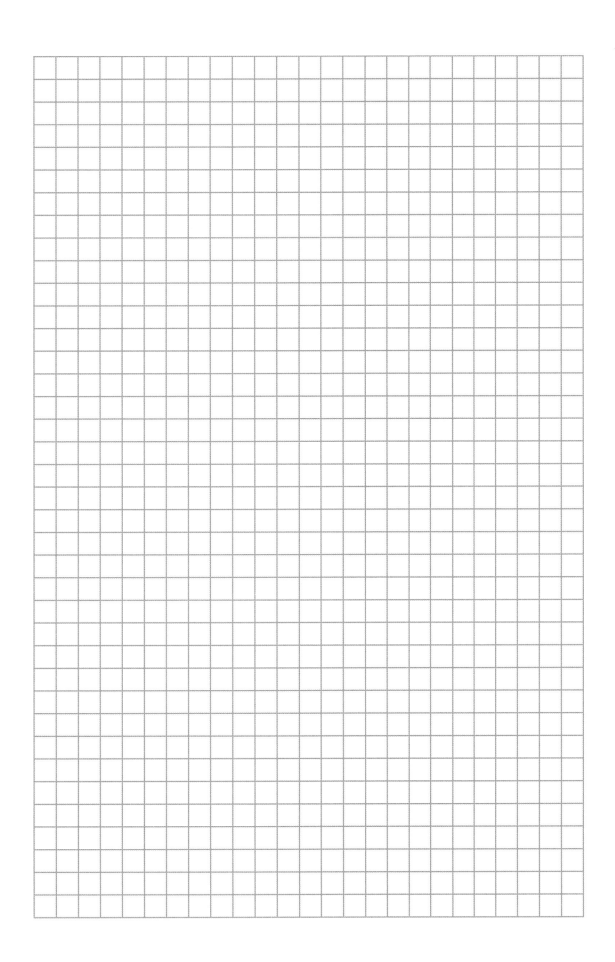

Acknowledgements

The authors and publisher are grateful to the copyright holders for permission to use quoted materials and images.

All images are © HarperCollins*Publishers* Limited

Every effort has been made to trace copyright holders and obtain their permission for the use of copyright material. The authors and publisher will gladly receive information enabling them to rectify any error or omission in subsequent editions. All facts are correct at time of going to press.

Published by Collins
An imprint of HarperCollins*Publishers*
1 London Bridge Street
London SE1 9GF

HarperCollins*Publishers*
Macken House, 39/40 Mayor Street Upper, Dublin 1, D01 C9W8, Ireland

© HarperCollins*Publishers* Limited 2022

ISBN 9780008551452

First published 2022

10 9 8 7 6 5 4

British Library Cataloguing in Publication Data.

A CIP record of this book is available from the British Library.

Publisher: Clare Souza
Authors: Samya Abdullah, Rebecca Evans, Trevor Senior and Gillian Spragg
Videos: Anne Stothers
Project Management: Richard Toms
Cover Design: Kevin Robbins and Sarah Duxbury
Inside Concept Design: Sarah Duxbury and Paul Oates
Text Design and Layout: Jouve India Private Limited
Production: Emma Wood
Printed in India by Multivista Global Pvt. Ltd.

This book is produced from independently certified FSC™ paper to ensure responsible forest management.

For more information visit: www.harpercollins.co.uk/green